P9-CCD-007

CREATURES OF ACCIDENT

CREATURES OF ACCIDENT

THE RISE OF THE ANIMAL KINGDOM

WALLACE ARTHUR

HILL AND WANG

A division of Farrar, Straus and Giroux

NEW YORK

HILL AND WANG
A division of Farrar, Straus and Giroux
19 Union Square West, New York 10003

Distributed in Canada by Douglas & McIntyre Ltd.
Printed in the United States of America
Published in 2006 by Hill and Wang
First paperback edition, 2007

The Library of Congress has cataloged the hardcover edition as follows:
Arthur, Wallace.
Creatures of accident : the rise of the animal kingdom / Wallace Arthur.— 1st ed.
p. cm.
Includes bibliographical references and index.
ISBN-13: 978-0-8090-4321-7 (hardcover : alk. paper)
ISBN-10: 0-8090-4321-1 (hardcover : alk. paper)
1. Natural selection. 2. Evolution (Biology). 3. Mutation (Biology). I. Title.

QH375.A78 2006
591.3′8—dc22

2005033540

Paperback ISBN-13: 978-0-8090-3701-8
Paperback ISBN-10: 0-8090-3701-7

Designed by Gretchen Achilles

www.fsgbooks.com

1 3 5 7 9 10 8 6 4 2

Such creatures of accident are we,

liable to a thousand deaths before we are born.

—Mary Antin, THE PROMISED LAND, 1912

CONTENTS

Preface ix

ONE *Hand Luggage Only* 3
TWO *The Biased Buddhist* 15
THREE *Sandcastles and Their Children* 27
FOUR *A Deafening Silence* 40
FIVE *Cork Prisons* 55
SIX *Building a Castle of Cells* 67
SEVEN *Dances with Genes* 79
EIGHT *Life Through Time* 89
NINE *The Embryo Wars* 101
TEN *Darwin and His Legacy* 114
ELEVEN *Possible Creatures, Probable Creatures* 126
TWELVE *Duplicate and Diversify* 136
THIRTEEN *The Best Box of All* 150
FOURTEEN *From Simple to Complex* 161
FIFTEEN *From Complex to Even More Complex* 172
SIXTEEN *Acquiring Your Head* 183
SEVENTEEN *Crossing the Threshold* 194
EIGHTEEN *Dinosaur Blues* 205
NINETEEN *Beyond Pluto* 216
TWENTY *Big Questions* 225

Glossary 233
Further Reading 241
Acknowledgments 243
Index 245

PREFACE

WE humans are the most advanced creatures on the planet. And since no other life-bearing planets have yet been discovered, we are the most advanced creatures in the known universe. We are conscious. We can think. We can ask questions. Perhaps the most intriguing of our questions is: Where did we and other complex creatures come from? How did humans, our fellow apes, and our intelligent cousins the dolphin and the octopus arise from simple unconscious beginnings?

There are two approaches to finding an answer to this question. One is to be open-minded: to use our brains to gather and analyze all the evidence we can find around us in an honest attempt to decipher the truth. The other is to start from some fixed premise like the existence of a Creator, and refuse to accept any answer that threatens this faith.

I am convinced that the open-minded approach is the only way forward. What could be more reprehensible to a benevolent Creator than blind faith, and its intolerant and often violent consequences? If there is a malevolent Creator, we are all doomed anyway. So, whether a Creator is benevolent, malevolent, or nonexistent, we should proceed to use our brains and analyze the evidence around us. That way lies possible enlightenment; the other way lies fundamentalism, superstition, and ignorance.

The evidence tells us clearly that complex creatures arose from simple beginnings by a process of evolution. And it also tells us that evolution is powered by natural selection, as proposed by those two great nineteenth-century naturalists, Charles Darwin and Alfred Russel Wallace. It is not guided by the ever-present hand of an invisible designer.

But natural selection alone does not explain how complex creatures can arise from simple ones. We need more. Most examples of natural selection in action involve the evolution of creatures that are only slightly different from, and no more complex than, their ancestors. The famous example of moths evolving darker pigmentation following the blackening of tree trunks by the soot of the industrial revolution is a case in point.

Understanding the evolution of complexity—ape from amoeba—is a far greater challenge than understanding the evolution of diversity *within* any particular complexity level: moths of many different colors from an ancestral moth of a particular color. The question of how animal complexity arose is the subject of this book. The answer has many facets. A particularly important one is the divergence of replicated parts. The parts can be big ones, like legs; middle-sized ones, like brain cells; or small ones, like genes. In all cases, the key principle is that when you have more than you need of something, you can afford to use your extra copies to create new things. And the acquiring of multiple copies in the first place was, as we shall see, just an accident.

This principle is by no means new. But it has been seriously underplayed, especially in the popular literature on evolution. My task here is to rectify this situation. In doing so, I hope to demonstrate to a wide audience that the rise of complex creatures can be explained with ease, but also with awe, by science.

CREATURES OF ACCIDENT

CHAPTER ONE

HAND LUGGAGE ONLY

Human thinking about the nature of life has been constrained by dogmatic ideological stances, not just religious and political ones, but also adherence to the scientific orthodoxy of the day. Such "philosophical baggage" needs to be jettisoned at the outset of our journey so that we open our minds to the maximum range of possibilities. Science should be all about open-mindedness and questioning. It is an honest search for the truth about the nature of "life, the universe and everything," to use that famous phrase introduced by Douglas Adams in his book The Hitchhiker's Guide to the Galaxy.

THIS book, like the animal kingdom, had an accidental origin. The seed from which it eventually germinated was sown by a newspaper article I read a few years ago. The retention of this seed of information for several years is a bit odd, because newspapers, unlike books, are ephemeral things. If you have already read today's paper, how many articles can you remember? What about yesterday's? Or last week's? We pass our time on trains and planes perusing the papers, but we don't commit much to memory—or at least to the long-term version of that everyday miracle.

However, occasionally something sticks. The article whose message stuck with me was in the color supplement of one of the Sunday papers. It was about evolution. A particular sentence is all that I retain, and doubtless in imperfect form. But, to an approximation, here it is: "Complex creatures, like humans, are a mere epiphenomenon in the history of Life."

What did its author mean by describing us as an "epiphenomenon," that is, a sort of blip on the periphery of something altogether more important? What was the deep philosophical point he was trying to impart? Well, his point was something like this. Since life began about four billion years ago, the vast majority of the creatures that have lived out their lives in every corner of our planetary home have been bacteria. This in itself is hardly surprising, because bacteria are very small, and small creatures tend to be much more numerous than large ones. You and I are both individual humans, but we are also both roving vehicles transporting millions of bacteria from place to place—most of them in our guts and on our skin.

But the point goes further. Not only are there far more bacteria than animals or plants; there are also more different *types* of bacteria. In other words, species. We all recognize different species of animals, whether very different, like humans and houseflies, or only slightly different, like horses and donkeys. In contrast, different species of bacteria require more than the naked eye to see, let alone distinguish. Indeed, the whole concept of a species—bounded by its members' inability to breed with other than same-species partners—is hard to apply in the bacterial realm, where reproduction is hardly sex as we know it.

Such difficulties aside, it is probably true that both now and at all other points in evolutionary time the living world has been dominated by bacteria, both in numbers of individual creatures and in numbers of types of creatures. This fact is at odds with a curious human practice—naming particular periods of the earth's history after a particular kind of animal, as in the Age of Fishes. Such names have a rationale but also serve to mislead. They are normally used to refer to a type of animal that diversified rapidly in the period concerned and consequently contributed much to its fossil record. But unlike fish, most bacteria have no hard parts and rarely fossilize. So fossil frequencies are a poor guide to the dominant creatures of the past. In reality, the whole of evolutionary time could be labeled the Age of Bacteria.

Under this view of life, we humans and our animal allies are indeed a "mere epiphenomenon," or, in more graphic terms, a tiny molehill on a vast bacterial lawn. A molehill that may one day disappear, whether by self-inflicted nuclear radiation or other means, leaving our simpler but more robust progenitors to go about their bacterial business in a molehill-free manner, as they were doing three billion years ago.

I'm going to call this lawn-with-molehills perspective the left-wing view of life. The reason for this label is that, in this perspective, attention is focused on diversity—that is, creatures being merely different from each other *within* a level of complexity—rather than on complexity, and on evolutionary *increases* in this over time. There is a reluctance to think of any local corner of this variety as being better or higher in some sense than another. So it is an egalitarian view of life. We life-forms are all comrades in our struggle for existence, whether we are bacteria, beetles, blue whales, or bus drivers.

So what is the right-wing view? This is where we go from lawn to ladder. A long time ago in Germany—about two centuries back—a group of philosophically minded folk interested in the nature of life came up with the idea that all creatures could be arranged in a line. And the line was a vertical one: it was effectively a sequence of increasing complexity and sophistication. Microbes at the bottom, humans at the top. This is something of a caricature, but it captures these philosophers' fundamental point.

This vertical line needs a name. It is often referred to by the Latin *scala naturae*, meaning, as you might expect, "natural scale." Whatever we call it, any overall view of life based on it is fundamentally at odds with the lawn-with-molehills view. If our central metaphor shifts from horizontal to vertical, we change our focus from diversity to complexity. From humans as molehill to pinnacle. From bacteria as all-dominant life-forms to lowly dwellers on the bottom rung.

Now, each of these views captures an element of truth. It is always unwise to regard thinkers of the past as ignorant or deluded. The further back we go, the less the information at the disposal of the thinkers concerned. But while lack of some kinds of information constrains thinking, it does not necessarily distort it. That is, it does not prevent a rigorous analysis of whatever information was available at the time. So I am not interested in being critical of last decade's Sunday newspapers or of worthy tomes written in the eighteenth and nineteenth centuries in a language that regrettably I cannot read. Sure, my criticism is implicit, but I will not develop it. Instead, what I will develop is a Middle Way, to borrow that famous Buddhist concept, between the left-wing and the right-wing views—that is, between life's lawn and life's ladder. However, it will be a decidedly biased Middle Way—more on that in the following chapter.

Today, the worldview of most Europeans, and many (but not most?) Americans, is sufficiently infused with evolutionary thinking that the idea of evolution in general causes an adverse response only in the creationist movement. There are now many strands of this movement, the most recent being intelligent design, or ID.

If all that phrase implied were an intelligent Creator causing the big bang through which the universe was born some fourteen billion years ago and then standing back to let nature run its course, it would seem no great threat to an evolutionary worldview. But unfortunately it goes further. I will return to this point toward the end of the book (in Chapter 20). Most of the book is not devoted to attacking ID, but rather to building the case that ID is unnecessary by explaining, in a way that is readily accessible to a general readership, how the rise of complex creatures can be explained in terms of natural processes. I will take the view that we should all accept evolution regardless of whether we do or do not see religious implications, because of the mass of accumulated evidence in its favor, some of which we will see in subsequent chapters. But accepting evolution was not so easy in the late eighteenth century.

We need to be careful here. What exactly was the state of human thinking 220 years ago? How uniform or varied were people's views of life? It is easy to make ill-informed statements about such times that historians of science could disprove in a single sweep of the pen. Belief in evolution did not begin with Charles Darwin and his 1859 masterpiece, *On the Origin of Species*. That book may indeed have been the biggest single milestone along the way from a dominance of creationist myths to the triumph of evolutionary arguments, but it was not the first. European views on evolution can be traced back at least half a century earlier, and probably further. And who knows what the Chinese were thinking about life way back centuries before Darwin's time, when they scooped Copernicus with their early ideas on a heliocentric solar system?

But caution should not go too far. Although you and I have little insight into the philosophical conversations that took place in late-eighteenth-century Germany, we can be fairly sure that evolution was not center stage at that time. I'd guess that some folk had thought of it. Regardless of whether this is true, evolution still did not feature in the prevailing European worldview by 1800.

This fact raises an interesting question. What was in the minds of those late-eighteenth-century proponents of the right-wing view? If life had a ladder, did creatures not climb up it? No, they emphatically did not. The ladder, which now so readily permits an evolutionary interpretation, was at the time very abstract—a pattern in the mind of God.

Now, in post-Darwinian times, creatures most certainly are seen to climb the ladder—but only some of them. Many creatures have remained on the lawn of design simplicity ever since life began. That is, they have remained as single cells. Others have climbed ladders that led to greater complexity, in some cases to designs that involved many millions of cells. Exactly how they climbed the ladders we'll get to later, and clearly Darwinian natural selection is a part—though only that—of the answer. For now, the main thing to notice is that the ladder has been pluralized. There has not been a single escape route from the realm of the simple; rather, there have been several. Some led to greater heights than others. After all, even some bacteria have climbed a short ladder to what are called filamentous designs. These involve cells sticking together in strings rather than all existing as entirely independent creatures. And plants climbed a very different ladder from animals.

It gets worse. Evolution is more like the child's board game Snakes and Ladders than it is like ladders alone. Once you have ascended far enough, it is possible to go down as well as up. As one line of ancestors and descendants (a lineage) is going up, another is going down. Now we get into a terminological and pictorial morass. Darwin called one of his books *The Descent of Man*. More recently, the Polish-British-American scientist Jacob Bronowski wrote a book, and presented a TV series, called *The Ascent of Man*. In biological texts, evolutionary trees are usually pictured as going up; but occasionally they are shown as going down, or even sideways. Human family trees, and diagrams of genetic crosses, are usually shown as going down, though they are mere microcosms of those bigger evolutionary trees that usually go up.

The morass in which we now find ourselves is sometimes referred to as philosophical baggage. Every individual thinker carries into every stage of his or her thinking a set of views (good) or prejudices (bad) that have been born of previous reading, thinking, and talking. We are by nature reflective creatures, albeit some more so than others, and we cannot help forming pictures of the world as we learn about it. But these pictures, although they are essential tools in the learning process, can lodge too firmly in our minds at too early a stage and inhibit shifts to different pictures that may become appropriate as we learn more. And this problem characterizes the whole human learning endeavor at the community level as much as it does our own individual mental journeys through life. We must be eternally vigilant about the need to discard our philosophical baggage when it becomes a hindrance rather than a help. Nobody should underestimate the difficulty that lies in maintaining this constant vigil, and in using it to good effect.

A particular kind of philosophical baggage lies at the heart of the morass of ascent and descent metaphors that we have just reached. This baggage concerns our particularly human identity crisis: our views, which tend to be strongly held because they matter so much to us, about what kind of creatures we are, how we got here, where we are going, and whether there is any ultimate purpose in our journey, save that which we make up ourselves as we travel.

This is the worst kind of baggage because it is the most important and thus the most difficult to discard. But discard it we must, even if only temporarily, in order to reach an unprejudiced view of the nature of life, a view based on facts rather than on ideology. The first step in discarding it is to inspect it carefully so that we can see its full extent. You can't discard baggage that you don't know you're carrying.

So back to the left-wing and right-wing views of life, or the lawn and the ladder. This time I'll take them in the opposite order, which is the order in which they historically arose.

The right-wing view emphasizes differences rather than similarities. Moreover, it emphasizes a single dimension along which (or, more accurately, up and down which) the differences among creatures are manifest. But what is this dimension? I referred to it earlier as being complexity or sophistication. Other terms could be, and have been, used. Some creatures can be described as of a higher grade or as being more advanced than their lower or more primitive relations. It is probably reasonable to describe humans as being more advanced in some sense than flies. Is it equally reasonable to describe flies as being more advanced than snails? Probably not—they're just different. What if we replace snails with their brighter molluskan cousin the octopus? Most people would describe the octopus as more advanced than the fly, largely because of the octopus's greater capacity for learning.

Notice, however, that learning is not the same as complexity, though they are related. It is probably true to say that no very simple animal has a large capacity for learned, as opposed to instinctive, behavior. There are no brainy flatworms. But there certainly are many animals that are structurally complex yet limited in the realm of learning. This is true, for example, of sea urchins and centipedes.

Worms lead us to another problem. Not only are structural complexity and behavioral complexity different, but complexity of life cycle is different again. Some parasitic worms, most famously tapeworms, are structurally quite simple as adults but exceedingly complicated in terms of the life cycles through which their adult forms are reached. These life cycles can consist of several stages that bear little resemblance to each other. The difference between a caterpillar and a butterfly springs to mind here. But some parasitic worms make the insect life cycle look simple, as they have more stages than just two—or three, if we count the butterfly's chrysalis as a stage.

The complexity of creatures should not condemn us to complexity of thinking about them. So let's now try to emerge from our linguistic confusion. Paradoxically, complexity is simple to define. A commonly used definition of the complexity of a creature is its number of different types of constituent parts. But as ever, the devil is in the detail, and in this case the detail concerns the nature of the parts. Life stages, like larvae, are parts of life cycles. Organs, like hearts, are parts of organisms—at least of some organisms at some stages of their life cycles. And, at a lower level than organs, cells are also parts of organisms.

Different axes of complexity run in parallel to varying degrees. Creatures with a greater range of cell types will often have a greater range of internal organs, simply because, for example, it takes different types of cells to make a brain and a heart. But creatures with more organs do not necessarily also have more life stages, as we have just seen. What this means is that in a comparison between two creatures, it is sometimes possible to say in a general way that one is more complex than another. Humans, for example, are more complex (by far) than bacteria. However, we sometimes have to be more specific, and say that creature X is more complex in some ways than creature Y, but simpler in others.

Now to the left-wing, or lawn, perspective. Here the vertical axis, whatever its precise nature, is downplayed. Instead of emphasizing that some creatures are higher than others on this axis, it emphasizes that most creatures have retained a very simple body form—the unicellular one—but have diversified within this constraint. That is, they have become very different from each other, but without any of the different forms being significantly more complex than any others.

Given that both lawn and ladder views capture elements of the truth, how can we best picture the evolutionary process? Think of a three-dimensional space, such as a room. Let's imagine that complexity starts at the floor and heads for the ceiling. The length and breadth of the room can be thought of as two variables, each representing some aspect of being different without differing in complexity. Three billion, two billion, and perhaps even one billion years ago, our room of life was carpeted but unfurnished. Or, if you prefer not to multiply the metaphors, it had an indoor lawn. There were no complex creatures of any great height on the vertical axis. Today's world, however, is very different. The carpet (or lawn) is still there. It may even, paradoxically to anyone familiar with real carpets, be deeper-pile and have fewer holes than before. But the room is now characterized by multilevel furniture. It contains everything from footstools to bookshelves. The vertebrates, arthropods, mollusks, and others stand tall, but they don't outnumber the fibers of the bacterial carpet.

Problems are melting away fast. A Middle Way between lawn and ladder views is not only possible but necessary. It is a more accurate view of how our biosphere came to be populated by the creatures we see around us. And in terms of representation, it doesn't matter whether we choose to look at our room the right way up, or upside down, or even lying on its side. Nor does it matter that different measures of complexity will result in different arrays of furniture—indeed this is to be expected.

But a problem still remains. Where has time gone? We have pictured two rooms separated by at least a billion years. In such a comparison, time is represented only by the contrast between two (or more) rooms; it is not represented at all within any one of them. Recall that "up" means more complex. Therefore, although there may be a tendency to think of "up" as also meaning later, this cannot be true. If particular lineages can go up or down (or stand still) in complexity terms as time proceeds, then an axis measuring complexity cannot also be an axis of time.

The conflation of time and some measure of complexity, advancedness, or grade, regardless of what exact form this takes, has in my view been a major contributor to unwanted philosophical baggage in much thinking about the evolutionary process. And since evolutionary trees are usually drawn on pages, posters, or screens with a mere two dimensions, as opposed to our imaginary room of life with three, the problem is even more acute. The horizontal axis becomes "degree of difference within a level of complexity," the vertical one simultaneously grade and time, thus producing the false notion that evolution is an ever-upward process.

An important step in solving this remaining problem was made by the German biologist Willi Hennig in the mid to late twentieth century. Now, Hennig is not a household name in the way Darwin and Wallace, or Watson and Crick, are. But perhaps it should be. For the advance that Hennig made is one of those brilliant, obvious-in-retrospect ideas. He sought to separate two things: (*a*) the connectedness of ancestor-descendant lineages in time, together with the splits that occur in such lineages; and (*b*) the evolutionary modifications (in terms of changes in the structure and function of the organisms concerned) that were happening in each lineage.

Hennig chose to concentrate on the former, so the vertical axis in his branching diagrams was time—not in an absolute sense but in the form of a temporal sequence of lineage splits. Picture a V based on one of the branch tips of another V to get an idea of one lineage divergence preceding another. His horizontal axis was only there to allow lineage splits to be depicted at all—something that can't be done if you only have a single axis (picture an I stacked on another I). The horizontal axis emphatically does not represent degree of difference, as it does in other kinds of evolutionary trees.

If we return to imaginary rooms, it becomes clear that we could design a better type of room than the one we were considering earlier. Effectively, the way it was designed wasted a dimension. Suppose we redesign it as follows. Change the vertical axis to "time." Change the length axis to "complexity." Retain the breadth axis as a single compound variable (for surely only one is needed) to refer to "changes within a level of complexity." Any lineage is now constrained to go up, but it can meander in both directions on the other two axes. Time and complexity can increase together or move in countercurrent; or complexity can stand still as time moves on. Strangely, I have never seen such a picture of evolution in print—but given the volume of the relevant literature, it may well be out there somewhere.

This book, like others, is a mental journey. If you and I, and our fellow travelers, are weighed down with too much philosophical baggage, such as an obsession with lawns because of the previous excesses of the ladder proponents, we will burn up most of our fuel just dragging this baggage along, and will reach a destination not far from where we started. But it needn't be like that. People cannot discard all their prejudices, whatever they may think. I don't ask the impossible of you, or indeed of myself. I don't ask that we travel naked as the day we were born. Rather, I ask for the flight to be restricted to those who can travel with (philosophical) hand luggage only. That way we at least have a fighting chance of flying to an exotic destination.

CHAPTER TWO

THE BIASED BUDDHIST

The Buddhist concept of a Middle Way is essentially an advocacy of the avoidance of extreme stances. So it is an argument for compromise. The strength of such an argument is context-dependent. In politics, compromise is often the best way forward. In science, sometimes it is, sometimes not. In this chapter I advocate what might be called a biased Middle Way for dealing with the evolution of complexity. Taking this approach, we acknowledge two dimensions: a vertical one of increasing complexity and a horizontal one of diversifying within a level of complexity. But we pay more attention to the former than to the latter.

FOR a long time, most people thought the earth was flat. On the basis of the evidence available in bygone days, which consisted of the view from the highest hill, flatness on a grand scale seemed the sensible conclusion. Although local undulations were apparent in the form of other hills, the big thinkers of the time could mentally discard these and focus more on the underlying surface from which all the observable hills arose. And this indeed would have seemed flat.

Now we know better, not because individually we have better powers of observation of the landscape, but rather because as a society we have amassed all kinds of evidence of roundness. Most vividly, we have pictures of the earth taken from space, which are rather hard to argue with. So this was not a case for compromise. If, at an early stage in the perhaps-the-world-is-round debate, the two sides had sat down in a civilized way around a table and politely agreed to compromise, this would have been a great mistake. It would not have aided the progress of science. Indeed, it is difficult to imagine what shape (literally) a compromise theory would have taken in this case. Perhaps a sort of slightly bulging disk?

Anyhow, the general point is clear: sometimes in science compromise is not the right strategy. It may be temporarily uncomfortable—even downright scary—to be involved in a fierce debate in which the scientific combatants receive a mauling of their personal reputations. A friendly handshake over a compromise might seem more attractive to some. But if we wish to advance human knowledge, we ought not to make decisions on the basis of personal comfort.

There have been many other scientific debates to which the same applies. Either the earth is at the center of the solar system, as used to be thought, or the sun is, as we now know. Again, it would have been wrong to compromise. And with regard to the fact of evolution, as opposed to any particular theories about how it happens, the same is true. Imagine the possible nature of compromises in these two cases. Perhaps a solar system in which half the planets revolve around the earth, the other half around the sun. And on the earth itself, perhaps some species created in immutable form by a divine being, the others appearing over the eons by evolution. What nonsensical ideas! But they are nevertheless useful because they take us straight to something that is at the heart of science—the search for simplicity.

The philosopher Alfred North Whitehead once said: "Seek simplicity but distrust it." This quotation, better than any other one of comparable length, captures the true nature of the scientific endeavor. We seek to explain as much as possible of the detail of life, the world, and the whole universe, with the simplest and fewest theories possible—and they need to be testable theories so that we can see if they are trustworthy. This principle turns up in various guises all over the place. If there are competing theories to explain phenomenon X, then, other things being equal, we choose the simpler one. This choice can be called the principle of parsimony. I suppose it could also be called being economical with our search for the truth (in contrast to being economical with the truth itself—a phrase that carries very different connotations). Sometimes, when a theory is too complicated, we try to cut it down to something simpler—a process known as wielding Occam's razor, named after the fourteenth-century English philosopher William of Occam.

Anyhow, back to the question of: To compromise or not to compromise? Although there are plenty of cases in which scientific compromise is a worthless fudge, there are also cases in which it is appropriate, even necessary.

In the eighteenth century, there was a vigorous debate about the nature of human (and other) embryos. Some folk said that each generation began with a minute preformed human lodged in either the sperm or the egg, and effectively it just grew. Others said that there was no such thing, and that instead each generation of embryos was formed from scratch.

There are two ways of looking at this old debate through modern eyes. One is that the preformationists were simply wrong, and the rival camp—referred to as the epigeneticists—right. The other, championed by the Canadian philosopher of science Jason Scott Robert, is that both views captured an element of the truth. Although embryos are not preformed, their development in each generation is based on something that is—the genes.

The conflicting lawn and ladder views of life also present an opportunity, perhaps even a need, for compromise—but only up to a point. I should explain what I mean by this slightly strange statement.

The author of the newspaper article I referred to at the outset was the famous American paleontologist Stephen Jay Gould, who was a deep thinker and a prolific and adept writer. As such, there is no way that Gould really thought that we humans constituted a mere molehill on the bacterial lawn. Rather, he described us as an "epiphenomenon" in order to be provocative. He wanted to make his readers see his point with maximum force, so he made it in extreme form. Whether we chose to agree with his ideas or rebel against them was probably not of great importance to him as he wrote. What *was* important to him was that his words did not go unnoticed. And for good or for bad—probably the latter—extreme pronouncements are less likely to be overlooked than their carefully worded, more balanced counterparts.

The "nature philosophers" of the late eighteenth century who saw a ladder rather than a lawn were probably not being deliberately provocative in their writing. After all, there was little PR, hype, or spin in those days compared with our own, though it would probably be a mistake to think there was none. Nevertheless, that does not necessarily mean they would be less reasonable than Gould if you or I were able to bend time and sit down with them to discuss their cherished theories over an ancient German beer. Although we can't be sure, I would guess that most of them would readily agree that the idea of a *single* natural scale is too simple.

So, a classic case for compromise. Life's variety can be characterized by two dimensions, one horizontal and one vertical. The organisms of planet Earth have become different from each other over the course of geological time in various ways. Some of these have resulted in the many descendant species of a single ancestral one being of comparable complexity, and thus on the same rung of a ladder. In other cases, some descendants have climbed up (or down) a ladder and so have become "higher" (or "lower") than their progenitors. We're back to the three-dimensional room of life.

This view of life is clearly a Middle Way between two extremes. And as such it is preferable to either. But I said earlier that this was an area of scientific thought in which a compromise is only desirable "up to a point." It's now time to dissect this phrase.

Extreme lawn and ladder views of life are really just the ends of a continuum of possible ways of looking at things. Let's think in terms of percentages in order to clarify the nature of this continuum, which we can consider as a horizontal line of fixed length with 0 percent at its left end (lawn) and 100 percent at its right end (ladder). The percentage then becomes the degree of emphasis given to differences in advancedness, grade, or complexity.

So those writing provocative newspaper articles adopting a worldview of a diversity of creatures that are "just different" may be at 0 percent on this scale. In Gould's case, he did at least acknowledge the molehill, and so perhaps he was at 5 percent. It's possible to imagine a whole range of views from here right up to the 100 percent that is often thought of (incorrectly, I believe) as the position of the old German nature philosophers.

The compromise point that I arrived at above, in which we see both diversity and complexity axes, would seem a classic 50 percent compromise. And in some ways so it is. The lawn and the ladder each has its own dimension; neither is downplayed relative to the other. Philosophical baggage has been cast aside, as I urged in Chapter 1, and we are free to think about the nature of the earth's biota without feeling a need to constrain its immense variation to fit one mold or another. So far so good.

But now we hit a major problem. How does any practicing scientist decide what is most interesting in his or her chosen field? "Interest" is a curious word. In financial circles, the prevalent use of this word means you could say that one company's shares were "more interesting" than another in the sense that money invested in one would produce more interest (that is, profit) than another. This usage seems a far cry from the use of the word "interesting" in science. But strangely, it is not. In both cases, it is the level of interest that determines our course of action. Stockbrokers will invest money in the most interesting shares. And scientists will pour their time and energy into pursuing the most interesting questions.

It's much harder to pin down the scientist's goal than the stockbroker's. After all, wealth is a unidimensional thing. My bank statement does not have to take the form of a complex multidimensional table. Rather, it is a simple list, reflecting the fact that my balance at any moment in time is a point on a single (vertical) scale whose ends are undefined but range from large minus numbers (if I have an overdraft provision) to large positive ones (actual, or, for most of us, only potential).

Imagine the equivalent of a bank statement for the activity of a scientist. What would it look like? Asking such questions immediately makes it clear that we are grappling with a tougher problem. Yet it is not insoluble. It may not be possible to quantify scientific interest in a precise numerical figure. But it most certainly *is* possible to acquire a good understanding of its nature.

To do this, we should probably start with applied science, even though it is not my center of attention in this book. So we can focus on a scientist who is trying to find a cure for cancer or to devise a biotechnological process for minimizing the harmful environmental effects of some waste product. In this kind of context, one particular avenue of inquiry might be described as more interesting than another because those working in the field suspect (for whatever reasons) it is more likely to solve the core problem being addressed.

But what of pure science? This book is about the kind of science that is not trying to achieve practical goals at all. Rather, the goal is an intellectual one: understanding the nature of life on Earth, and, more specifically, the evolutionary process through which it has come about. This is a biological equivalent of physical scientists' attempts to understand the nature of matter and energy, and their relationship, which is most succinctly put in Einstein's famous formula $E=mc^2$. Einstein was interested in the nature of the physical universe. He was not trying to invent a widget. The fact that the ultimate widget, in the form of the atomic bomb, uses his matter-energy relationship to devastating effect should be a good reminder that pure and applied science often interact. But still, their motives are very different, and so, therefore, are their interpretations of interest.

In pure science, the level of interest is strongly related to the level of generality of explanation. Someone pursuing a theory of evolution that is only applicable, say, to leeches will be thought of as following a less interesting program of research than someone pursuing a theory of evolution that is potentially applicable to all life-forms. Equally, someone investigating how genes interact with other molecules to make a slug embryo develop will be thought of as doing less interesting science (if the processes are slug-specific, which they may not be) than someone investigating general mechanisms of gene action that are potentially applicable to the embryos of all species, including our own.

Not only do pure and applied science interact, as illustrated by the atomic bomb, but they are less different from each other in overall ethos than they might appear from the picture I have painted above. This is because generality of explanation always contributes to the level of interest of a scientific problem, even when it is very much applied. If an oncologist (a cancer specialist) comes up with a cure for bladder cancer, or for breast cancer, that's great. But if another oncologist comes up with something that will destroy cancer cells of all types in all tissues, and so has discovered a *general* cure for cancer, that's even better.

What, then, is most interesting about the process of evolution through which all the creatures we see around us have been produced? If we base our answer on generality alone, it would probably be twofold, and something like this: first, the overall pattern of relatedness—the evolutionary tree—of the entire living world; second, the general mechanisms that have driven evolution over the eons. These are often referred to as the pattern and the process sides of evolution.

Now, back to the Buddhist concept of a Middle Way, but this time we'll think of it in relation to the percentage scale of lawnness-to-ladderness that I pictured above. Whether investigating pattern or process, an evolutionary biologist would, on the basis of the argument so far, be very sensible to sit at the center of the scale—in other words, the 50 percent point—giving equal weight to becoming more/less complex and to diversifying within a level of complexity. But I'm not going to do this. Rather, I'm going to take a ladder-biased Middle Way, and this will permeate the entire book.

Apparently against my own advice, I'm about to pick up something that might at first sight look like a piece of philosophical baggage. We've been remarkably free from it since discarding it in the previous chapter. The interest foci of solubility and generality that we have just been discussing are remarkably logical. They don't seem subjective. But unfortunately (or perhaps otherwise), they're not all we need in order to understand the overall scientific endeavor. There's something else.

The American ecologist Robert MacArthur once remarked that great science had to be "guided by a judgment, almost an instinct, for what is worth studying." What could be more subjective than an instinctive feeling? Yet I believe MacArthur was right. Anyone doing pure science should be striving for generality. But generality of what? In the examples I gave earlier, it was generality of life-form, with theories restricted to narrow groups like leeches or slugs being less interesting than theories that are potentially applicable to all creatures. That's fine as far as it goes.

But there is another issue here. Consider the number of species that inhabit the face of the earth at any one time, and then at another time a few million years later. In both cases the numbers will be very large. The chances of them being identical are almost nil. Assuming, then, that they are different, what is the *cause* of this difference? There are only two candidates: speciation (the origin of new species) and extinction (the disappearance of old ones). These are not competing hypotheses; rather, they are complementary parts of the cause. It is inconceivable that in any million-year-plus gap of time, there have not been multiple speciations and extinctions. One adds species, the other subtracts them, and the actual difference between times A and B is determined by their net effect.

Now, here's a deceptively simple question: Which is more important? It's deceptive because there are two very different ways of looking at it. First, we could think in purely numerical terms. If the number of species increases between time A and time B, then there is an excess of speciation events over extinction events, so speciation could be said to be more important. Following the same rationale, in times of declining overall species numbers, extinction could be said to be the more important of the two.

The second way of thinking about this issue of the relative importance of speciation and extinction relates to their degree of scientific interest. Which is more important for a general theory of evolution? The answer to this question can easily be arrived at by considering how Darwin would have fared if he had written a book titled *On the Extinction of Species*. This would hardly have carried the same weight as the book he actually did write, which, as I mentioned earlier, was called *On the Origin of Species* . . .

I can think of two rather different reasons for this contrast. First, if we are interested in how living things appeared, we are interested in origins: the origin of life; the origin of multicelled creatures from unicells; the origin of particular large groups of creatures, such as the vertebrates; and the origin of our specifically human characteristics, including our large brains, which we can uniquely use to think about abstract concepts like the nature of life or light, and write books about them. Understanding why species sometimes go extinct tells us very little about origins.

Second, the origins of species involve two things: change occurring in an evolutionary lineage, so that the descendants are different from their ancestors; and lineage splitting, wherein members of an ancestral species inhabiting distant parts of the world change in different directions, in terms of their structure and behavior, to the point where they can no longer interbreed. Although the latter is important in explaining how the number of species can increase, which is the direction we have approached from, the former was more important in challenging the prevalent nineteenth-century belief that species are immutable, and indeed it was this kind of change upon which Darwin concentrated.

So we are back to levels of interest. Speciation is in some sense more interesting than extinction. Even when extinction rates explode, as they have done several times in the history of the earth, most famously when the dinosaurs disappeared, they provide no direct information about origins.

Looked at in this way, Charles Darwin was not just pursuing generality of explanation, as any scientist should; he was also using his instinct to decide what it was that he should look for a general explanation of. What I want to do now is return to the ideas of life's lawn and life's ladders, and bring some scientific instinct to bear, both on the relative interest of these two and on the relative interest of going up and down the ladders.

And so, back to origins. Let's focus on the origin of one rather particular thing: the head. Early animals had no head; most present-day ones do have a head. So, somehow, heads originated—not all at once, of course, but in stages. From no head to rudimentary head to well-developed head to sophisticated head is definitely a series of steps up the ladder of complexity. Personally, I think such steps are of enormous importance. My scientific instinct tells me they are special, worthy of particularly intensive study. And not just heads, but other things that weren't there originally, too, like limbs and eyes.

As you will have noticed, I am veering well off course if I want to travel baggage-free. I am no longer centered on the 50 percent point of our imaginary scale of "ladderness"; I have moved up to an unspecified higher point—perhaps 75 percent, maybe even 85 or 95 percent. But not, for sure, 100 percent. That would be a serious mistake.

Now, you might think I have begun to defend the indefensible. I started off urging you to abandon philosophical baggage, but I am now urging you to trust to scientific intuition. Are these not merely the same thing hiding behind different names? Well, I would argue that they are not, but I accept that it's a close call. Let me explain.

Baggage is bad, not so much because it is subjective rather than objective, but more because it constrains freedom of thought. If you are ideologically wedded to something, be it communism, a fundamentalist religion, or the idea that evolution must always go in an upward direction toward ever-higher creatures, then this constrains your thinking. It means that there are some things you simply won't consider. They are threats to your overall worldview.

Scientific intuition is also subjective—it has that in common with philosophical baggage. But it need not be constraining at all. It comes in the form of "I have a feeling that X is more interesting than Y, but I could be wrong." It is firm enough to be used to decide upon a particular topic as deserving of attention, time, energy, even research funds, but it is tentative enough to be reconsidered if, after a certain amount of study, it looks less interesting than it did.

Scientific intuition thus does not lead us into extreme positions. The fact that I find upward changes in complexity to be most interesting does not lead me to make false claims about their importance. For example, it is probably not true that this kind of change outnumbers the others in frequency. My guess would be that most evolutionary changes, in most creatures, leave their level of complexity more or less the same. In this book, I emphasize ascents of life's ladders for reasons that have nothing to do with their frequency. Rather, my rationale is that focusing on these ascents helps us understand how complex life-forms such as ourselves arose from very simple beginnings.

CHAPTER THREE

SANDCASTLES AND THEIR CHILDREN

To get a clear view of something, we often need to step outside our usual perspective. Here we temporarily become Martians—members of the Mars Space Agency in particular—in order to clarify certain important features of life on Earth, and indeed perhaps of life in general. We contemplate a picture of a sandcastle that has appeared on our screen, sent by an Earth probe that has landed on a Scottish beach. This strange object certainly does not look like the result of a natural geological process. But is it a life-form or something made by one? To answer this question, we need to think about what features delineate life from nonlife.

IT'S amazing how the incredible can become the mundane. When Galileo Galilei was using crude proto-telescopes to observe Mars, sometime around 1600, he was doubtless both excited by the things he could see and frustrated by those he could not. If a time traveler from around 2000 had appeared and unfurled a large color picture of the Martian surface, showing individual boulders nestling among the reddish dust, Galileo would have been astounded. It would, as they say, have blown his mind.

Recently, as I was sitting idly people-watching in the astronomy section of a science museum, I noticed entire families walking casually past a large wall-mounted picture of the Martian landscape without so much as a sideways glance, on their way to something more hands-on and three-dimensional. The picture had become part of the wallpaper. It might as well have been actual wallpaper from the 1960s titled "red boulder pattern" and stemming from some designer's imagination rather than from a camera mounted on one of the Mars landers.

Now invert your mind-set. You are an earthling no longer. Instead, you are one of those fictional, intelligent, and strangely humanoid Martians that abound in the literature of science fiction. You are employed by an organization called MSA (the Mars Space Agency, of course). Your mission is to discover whether there is life on Earth. Let's suppose that the year is 1859 (the year Darwin published *On the Origin of Species*). So not only are we pre–space age, we are also pre–radio age. There are no telltale broadcasts passing through your Martian atmosphere giving you clues to the existence of earthlings.

So you send out an Earth probe. It takes off successfully, covers the millions of miles of space that it must, and enters Earth's orbit, where it remains undetected. Doors in its underbelly open and a tiny Earth lander, just a few centimeters long, is ejected. It spirals down to Earth and lands on a large beach in the far north of Scotland. A mini-camera whirs out and takes a digital picture of a vast expanse of sand. The tide is out, and the camera, although pointed seaward, is only a few centimeters above the sand, and so can see no water. There is not a man-made object anywhere in the field of view. Nor are there any plants or animals visible. Just sand.

I doubt if this image will cause much excitement when it appears on-screen at MSA headquarters. No evidence of life—apparently. In order to maintain this illusion (for that of course is what it is), we'll need to make three assumptions: first, that the camera ceases functioning just after sending back its first picture, so it never sees the incoming tide; second, that the biochemical analysis equipment fails from the outset, so the microscopic creatures lurking among the sand grains are never detected; and third, that the month is February. The reason for this third, and at first sight rather odd, assumption of a winter landing is that human activity on the beach is likely to be limited, even nonexistent.

Now imagine a second probe, sent a few months later, which lands in July 1859. Although it's highly improbable, we'll imagine that this probe suffers from the same malfunctions as its predecessor. So again, all we get is a single photo. But this time, when it appears on-screen at MSA, there is a stunned silence followed by a cacophony of sound—shouting, whooping, cheering!

The cause of the excitement is something that to earthlings has become rather mundane, but to Martians is quite incredible (a reversal of the transition noted at the start of this chapter). It's a sandcastle—a well-crafted one complete with square towers and walls with parapets. Now, I have to admit that I don't know when the building of sandcastles by humans began, except that it presumably was predated by the building of real castles, of which there are several examples, in the far north of Scotland where the Martian probe landed, dating back hundreds of years. So by 1859, sandcastle building was probably almost as well developed an art as it is today. (My children have built some fine ones on those selfsame northern Scottish beaches.)

It may seem a silly question, but why all the excitement at MSA HQ? After all, a sandcastle is not a life-form, and it is life that the Martians are searching for. And although we humans know that sandcastles are *built* by life-forms, there is no reason why the Martians should know this as well. Perhaps Martian history never went through a castle phase, so what to us is a sandcastle is to them just an inscrutable shape on the beach. Why should this cause more interest than a boulder?

It's all to do with shapes. But we have to be careful here. Angularity, as found in castles, seems to be a rare result of natural processes such as erosion, where the more likely result by far is an irregular and rounded shape, like a boulder. (Of course, subterranean Martian children may occasionally come up to the surface and make dust boulders, in which case our own interpretation of our photos of the surface of Mars has been seriously incorrect.) So the angularity of the sandcastle seen on the screen at MSA is promising. But there is an issue of scale here. Crystals are regular and angular; so are snowflakes. Both are entirely natural and nonbiological. So neither regularity nor angularity is uniquely found in objects fabricated by life-forms. But perhaps these qualities are unique in this way when they occur at an appreciable size—like a meter across. If the Martians know this, then their excitement at seeing a sandcastle is entirely justified.

Now along comes a big question for them. Is the sandcastle in the photograph a life-form or a product of a life-form? Resist, if you can, the notion that the answer should be obvious. This is only the second-ever Martian Earth probe and photo. The Martians know almost nothing of our planet. The sandcastle doesn't look like a nonliving boulder; it seems pregnant with information about life. But what should cause it to be interpreted as something that is *not* a life-form itself?

Regularity is not the answer, as most creatures here on Earth (and elsewhere, too, I'd be prepared to bet) are regular in many ways. Most animals, for example, are bilaterally symmetrical, meaning that the left side is an approximate mirror image of the right side. Nor is angularity the answer. Although most creatures are bounded by gentle curves, there are exceptions. Think of the sharp points at the ends of a stickleback's spines, or the almost perfect triangle that can be seen when the beaks of some birds are silhouetted against the evening sky.

But maybe rectilinearity (that is, not just having angles, but having right angles) is a more reliable pointer. Not only can I not think of a geological process that produces meter-scale rectilinearity; I cannot think of any animal or plant that takes this form, though some come close—for example, the box jellyfish. So if this is true of life in general, and not just of life on Earth, then the shape of the sandcastle may indeed point strongly to its being a product of life rather than being a life-form itself. Certainly, if we consider the range of human products that are rectilinear, it is enormous: bricks, cardboard boxes, doors, and (approximately) books are among the many objects that spring to mind.

So when the inevitable press conference begins at MSA headquarters, the spokesbeing announces with confidence that they have found clear evidence of an entity made by an earthling life-form—probably an intelligent one, but let's not leap to conclusions here; think of a wasp's nest in cross section for impressive angularity albeit not in the form of right angles.

Although this is a reasonable interpretation of the photographed sandcastle on the basis of its shape and size, there are other aspects of sandcastles that would be even more diagnostic in terms of life-form versus life-product. These emerge when we take a time-extended view of sandcastles. Our hypothetical Martians can't do this, as their photo has only captured a single instant in time, so let's ditch the Martian perspective and think about sandcastles as we better-informed humans know them. In other words, let's do a bit of sandcastle-ology.

This is where the children of my chapter title come in. We can imagine two types of children that a sandcastle could have: those that make it and those that it makes, respectively real and fictional. (I'll ignore here the fact that some sandcastles are made by adults.) When children make a sandcastle, a mathematician might summarize the process in this formula: children + implements + sand → sandcastle. Certainly, the children, their plastic spades, and the sand all come first; the castle comes next.

What happens after that? Well, effectively, not much. The children go home, the sandcastle remains inert, the tide comes in, the castle is gone. End of story. The sandcastle itself makes nothing. This comes as no surprise to an earthling, but a Martian using only a single photo and some scientific intuition cannot afford to jump to such a hasty conclusion.

Let's consider another possibility: sandcastle as life-form. If a sandcastle were indeed alive, what features would we expect it to have? Actually, very few features are diagnostic of life in the sense that they are found in all life-forms but nowhere else. The English evolutionary biologist John Maynard Smith, who died recently, gave a list of "life essentials" in one of his books, and it was only about four items long. The most obvious item, and the one I want to focus on here, is reproduction.

All kinds of creatures have the ability to reproduce. Not all individual creatures actually do reproduce, as is clear to us even from consideration only of our fellow humans, but every species has this ability, and, because enough of its member organisms achieve reproduction in practice, the species endures. Eventually, all species become extinct because at some point this rule ceases to be true and the level of reproduction is no longer sufficient to balance the mortality rate—but we'll ignore extinctions for now.

So an interesting question for the Martian scientists is: Will this thing (perhaps they will give it a name, but it will not be "sandcastle") breed? Other questions are also interesting, but less so given our examination of the meaning of scientific interest in the previous chapter. An example would be: Will the sandcastle move? From a Martian perspective, there is no reason why it should not have lots of tiny feet that are invisible because they are exclusively on its underside—a bit like the tiny tube-feet found on the undersides of starfish (or sea stars, if you prefer) that enable them to glide sideways across the substrate.

There's a bit of asymmetrical reasoning here, as follows. If the sandcastle moved sideways across the beach unaided, and retained its structural integrity as it did so, you and I would be more surprised than the Martians. They would probably conclude that the sandcastle was a life-form, while we might begin to inquire about our own sanity.

However interesting the question of mobility is, it is less interesting than reproduction in the context of delineating the living and the nonliving worlds because many organisms are static and many nonbiological entities move. And it is not just the plant kingdom where the combination of life and stasis prevails. If a Mars-launched Earth probe landed on a rocky shore rather than a sandy one, its camera might whir out and take a picture of the rock facing it, which is covered in barnacles. And if the camera managed to remain functional for a while so that it could take a series of time-lapse photos, it would not reveal barnacles gliding slowly around. After all, barnacles, when they become adults, are effectively welded to the rock, so it is hardly surprising that they do not move. (Limpets, although they look equally immobile to us, are very different and could be caught gliding slowly along by a series of photos.)

So, many living creatures are immobile. Equally, many nonliving things move. In a strong enough wind, the beach itself will move—a sandstorm. But this does not mean it is alive. Thus the two pairs of opposites life/nonlife and moving/static do not align in a neat way. This means that observations about movement or its lack are not necessarily informative about the nature of unknown objects on distant planets.

But reproduction is, as they say, another kettle of fish. Not only are there no kinds of creatures that are incapable of reproduction (at least when "kinds" means species, so we can avoid complications like nonbreeding worker bees), but also there are—I think—no nonliving entities, either natural or human-made, that *do* reproduce. Boulders do not have children; neither do cardboard boxes. And neither do sandcastles.

You might want to question this broad generalization, especially for some of the products of today's human society. It's clear that one cardboard box can't make another. Rather, both are made by a cardboard box factory—probably a very automated one these days. Such a factory is very good at making boxes, but it is entirely incapable of making another factory. So far so good. And I think that if you cast your thoughts across all sorts of other human fabrications, you will find a similar story: we make things that can make other things; we do not make things that can make more of the same things—in other words, can reproduce. I keep saying "I think" here because there's something niggling away at the back of my brain telling me that there may be exceptions, particularly in the world of computers. Not hardware, software. A computer virus is called just that because it is capable of spreading in a way that involves a form of reproduction. So we might have to be a little more careful and say that no nonliving *real, three-dimensional* entities are capable of unaided reproduction. But this restriction is fine. There are, as far as we know, no life-forms that exist only in cyberspace. In the 3-D world, then, the pairs of opposites reproducing/not-reproducing and living/nonliving align perfectly.

It is this perfection, or *generality*, of relationship that makes reproduction more interesting than movement to a scientist—whether on Earth or on Mars—who is trying to decide whether a previously unseen object is alive. Lack of movement in a sandcastle is less informative than lack of reproduction.

Now let's ask a seemingly bizarre question that can be answered at several levels: Why do sandcastles not reproduce?

The most obvious answer—because they're not alive—is of little use to us here, as it takes us round in a circle. If we define life-forms as things that reproduce and then say that sandcastles don't give birth to baby sandcastles because they are not life-forms, we are guilty of circular reasoning—or of producing something called a tautology. So we have to go deeper and consider what it is about being alive that makes reproduction possible, or, conversely, what it is about being inanimate that takes reproduction out of the repertoire of typical behaviors.

Living things are made of cells. Nonliving things are made of a variety of different materials—metal, plastic, stone, and so on. Sandcastles, of course, are made of sand. There is something about cells that makes reproduction possible, and equally something about sand that rules it out. What are these somethings?

Let's examine the two building materials, sand and cells, in that order—sand right now; cells, after a brief digression, in Chapter 5. So here we ask: What is sand? Or even: What is a grain of sand? There is no single answer to this second question, because sand is a mixture of two (main) things. Some grains are minute particles of rock; others, often in the majority, are tiny fragments of the shells of long-dead mollusks. In other words, although sand is clearly not alive, and some sand grains have never been graced with a living source, others are very much the products of just such a source—and indeed have collectively constituted a sort of home for their source creature.

But strangely, this duality in the nature of sand doesn't matter. In making a shell, a mollusk is constructing something that is not itself alive. It is just as inert, especially after the death of its maker, as a stone. It is subject to weathering, like a stone. It can be moved around passively by the tide, like a stone. But it cannot move itself. A shell and a stone are to all intents and purposes identical, despite their contrasting origins. And when shells and stones are shattered and eroded into grains of sand, it matters not from whence an individual grain came.

The reason it doesn't matter is that, whether biogenic or otherwise, grains of sand are inert. They don't do anything. Their chemistry is that of the rocks. And if they can't do anything, then neither can those larger objects, such as sandcastles, that are made up of them. This is why sandcastles cannot reproduce.

A little caveat is in order here. If a grain of sand fragments into two parts, it has produced two sand grains. This fragmentation could be thought of as a kind of reproduction. But it is very different from its biological counterpart, and not just because the "offspring" are smaller than their "parent"—after all, baby humans are smaller than their mother and father. Rather, it is because there is no sand-grain counterpart of growth. Nor, of course, is there a sand-grain equivalent of sex, though not all reproduction in the biological realm is sexual, as any gardener knows.

Typically, grains of sand are bigger than cells, but the size ranges of the two overlap. The biggest cells—often egg cells—are bigger than the smallest grains of sand. Let's consider a cell and a grain of sand that are identical sizes (say, a millimeter in diameter) and similar shapes (say, irregular but approximately spherical). Imagine each blown up to ten thousand times its real size and shown in cross section on a huge screen. Now imagine that rather than remaining as a static image, each is turned into a movie. That is, we are able to watch changes over time.

To continue the movie metaphor: Imagine you are in a two-screen cinema. On screen 1, there is *The Story of a Sand Grain*. On screen 2, there is *The Story of a Cell*. It's a rainy day, so you decide to see both. Unwisely, you decide to watch *Sand Grain* first. To say that the plot is slow would be an understatement. Although you are watching a movie, nothing actually moves. If you stay long enough, perhaps a small chip will fall from one side of the sand grain's periphery. But that's about it. After half an hour, you admit defeat and head for screen 2, with some apprehension, thinking that *The Story of a Cell* may be no better.

It does not take you long to realize that your apprehension was unfounded. The two movies may have unimaginatively similar titles, but their plots could hardly have more different speeds. Watching screen 2, you sit mesmerized by the myriad movements of a cast of thousands. Molecules are whizzing everywhere. They are made, altered, moved, and ultimately destroyed by other molecules. There are multiple subplots all played out in parallel. There are even actors so small they cannot be seen at this resolution of ten thousand times reality—just as well since things are hard enough to follow as it is. As the movie reaches its climax, the huge orb of the cell writhes at a macroscale and, with the orchestra reaching a crescendo, splits down the middle. It has reproduced.

Reproduction of creatures is based on reproduction of cells. Indeed, in some cases—as with unicellular creatures like bacteria—the two processes are the same. In other cases, as with us humans, who each consists of many billions of cells, the connection is still there, even though the scales of cell reproduction and organism reproduction have become different by many orders of magnitude.

I'll continue the story of cells in Chapter 5. But before leaving the present chapter, I'd like to approach the creature-versus-sandcastle contrast from one final angle. So far, we have been content to believe that creatures reproduce because they are made of cells, while sandcastles do not because they are made only of inert sand grains. We have seen the origins of grains of sand in the fragmentation and erosion of bits of rock and molluskan shells. But we have not considered the origin of cells except in its shortest-term sense—one cell as the parent of another.

If we trace one of our own brain cells back in time, we arrive, ultimately, at our individual origin as a fertilized egg. If we go back further, we can span as many generations as we like. The further back we go, the less familiar will be the creatures concerned. Eventually, we end up with an ancestral unicell. But suppose we keep on going back. What then? After countless generations of unicells, where will we end up? Probably with ill-defined proto-cells that are much less sophisticated than any of their present-day descendants. Earlier still, we would find mere aggregations of large molecules that have "learned" to reproduce—a self-sustaining population of croutons in the primordial soup.

Of course, molecules cannot learn in the sense that we normally understand. It is not that particular molecules were "trying" to reproduce. It was just that some did and some did not. The latter perished while the former prevailed. This process is micro-level Darwinism, and it is how the cell, and thus ultimately all creatures, began. The rest, as they say, is history. More reliable reproducers outlived others.

This approach leads to a different answer (from the cell-based one) to the question of why sandcastles do not have children. They are not the products of a long evolutionary process in which those entities that produced more children than others, and did so in a more reliable way, survived, while other, less efficient reproducers went to the wall.

Which is the better answer? Is it more satisfying to think of reproduction as a feature of entities made of cells, or of entities that are the product of countless generations of natural selection? In fact, these are just different ways of saying the same thing. Acknowledging this fact raises further questions, the most interesting of which is whether it is possible for an evolutionary process on another planet to produce reproducing life-forms made of units that we would not describe as cells. That's a question we cannot yet answer, but may be able to one day as a result of a "close encounter." Anyhow, the quest for a complete understanding of life on planet Earth is a daunting enough challenge, and one that will keep us supplied with interesting questions for a good while yet. So I will stick with terrestrial life until near the end of the book, after allowing myself the indulgence of one further brief alien speculation.

Is there a planet in some remote galaxy where living sandcastles made of cells give birth to baby sandcastles, on cellular beaches strewn with inert animal-shaped boulders made of compacted sand? My guess, or scientific intuition, if you prefer, is that no such scenario can be found beyond the realm of the imagination. But suppose for a moment that I am wrong. Suppose we eventually find such a planet. Which would we consider more interesting—the beach (= unicells) or the sandcastles (= multicells)? I'd say the latter, plus the transition between the two—that is, how the sandcastles evolved from the beach. So we're back to that biased Middle Way again. All life-forms are interesting, but some are more interesting than others.

CHAPTER FOUR

A DEAFENING SILENCE

As I said at the outset, it is not just religious and political views that can constrain our understanding of the nature of life. Science itself can also impede progress in our understanding, if an inaccurate or unbalanced view becomes too entrenched. In the case of the evolution of complexity, the problem is not inaccuracy but lack of balance. Too much attention has been focused on diversification within a level of complexity and not enough on changes, particularly ascents, in complexity. This imbalance is readily understandable. It arises from the fact that diversification happens more rapidly than complexification, and so is easier to study. But it has had an unfortunate consequence. Much of the evolutionary literature has seemed to ignore what can be regarded as the most important feature of all: the rise of advanced creatures from simple beginnings.

OVER the last half century or so—say, since the discovery of the structure of DNA by James Watson and Francis Crick in 1953—what evolutionary theory has had to say about the origin of complex creatures might be described as in the title of this chapter, namely a deafening silence. Inasmuch as silence means no sounds at all, either real ones, as in conference talks, or figurative ones, as in written publications, my statement is too sweeping. But I make it for the same reason that Stephen Jay Gould made his overly sweeping comment about humans being a mere epiphenomenon in the history of life. That is, I don't want the point to go unnoticed.

Assuming, then, that I have your attention, I can admit that there have been intermittent lone voices calling out in the night about the importance of complexity, but they have been so few, and so overwhelmed by the multitude of voices talking merely about diversification—or becoming different, if you prefer—that they have made little impact.

I can think of only a handful of books in the last fifty years that have had the evolution of complexity as their central theme. And in the last five years there have, to my knowledge, been none. Authors that deserve a brief mention as producing the handful include the American biologist John Tyler Bonner, and the English biologist John Maynard Smith, writing jointly with his Hungarian colleague Eörs Szathmáry.

So why the deafening quasi-silence? Why has a potentially major theme in evolutionary biology with huge philosophical significance been written about, and talked about, so little? This is the sort of question to which you can get a lot of different answers if you ask it of a lot of people. So take what follows as a personal view, and think about whether you might want to add some additional reasons of your own that I haven't covered. My own preferred reasons are twofold—one ideological, the other, in contrast, quite pragmatic. I'll take them in that order, and will devote more space to the latter, as I gave the former—in the form of the lawn and the ladder views of life—a good airing in Chapter 1, and it just needs a little more fleshing out here.

I think the nub of the ideological problem is that the ladder view of life has, perhaps even from its inception, been tainted by association. Or rather associations in the plural, as there have been both scientific and political ones. The old idea of a natural scale, under which all creatures were arranged in a vertical line with microbes at the bottom and humans at the top, has been reacted against by most biologists for both of these reasons. Scientifically, it is too simple—life is not linear in any sense, except, perhaps, for the linearity of time, which sends us in a straight line from birth to death. Life, in the sense of the entirety of creatures, has a more complex shape by far—hence the range of tree and bush metaphors in widespread use.

As we begin to stray from science into philosophy and politics, the plot thickens. There is something distinctly unsavory about considering ourselves the pinnacle of the evolutionary process. We may be tops in intelligence, but we are way down the league in other characteristics, such as size, strength, and longevity. Evolution has not one pinnacle but many. Furthermore, it is prudent to recall that evolution has not stopped. It may run on until the sun dies—an event that we know will be many millions of years in the future. What subsequent creatures it will produce, we as individual humans of the present will never see, unless through a one-way mirror from some speculative "scientific observer afterlife."

So the idea of a single natural scale of complexity and sophistication of creatures is unsavory, egocentric, and just downright wrong. Although these are reasons enough for the reaction against it, there is another. And this involves real politics rather than what you might call scientific politics.

One of the leading figures in evolutionary biology in the Darwinian era was a brilliant German biologist, Ernst Haeckel, who took a ladder-based view of life, like his predecessors half a century earlier. The main difference was that patterns which from the old nature philosophers' viewpoint were in the mind of God were for Haeckel mechanistically explicable through the nature of the evolutionary process. This was a big advance, of course, yet it came with some unwanted baggage.

There were two problems. First, Haeckel was (and still is) seen by many biologists as being totally right-wing in the scientific sense, and supporting the old idea of a natural scale. I think this perception of his work is wrong, but that's another story in itself and I'll leave it until Chapter 9. Second, Haeckel's view of the human species was racist: he regarded Caucasians as superior to all other racial groups.

When a leading proponent of a particular scientific view is seen as a flawed character, and particularly one who holds unsavory political views, this perception has a major effect on the behavior of subsequent scientists in the field. It is not good to be seen as a disciple of such a person. And this kind of effect can be very long-lived—it can last for decades, even centuries.

My argument here is that we should try to resist the whole idea of a major scientific concept being tainted by association. We should judge any concept on its own merits and not on the basis of the personality or politics of its originator or its most prominent advocates. And we should avoid writing something off in total just because its most extreme form is clearly wrong. I have no patience for an extreme right-wing view of life that arranges all creatures in a line. But equally, I have no patience for the extreme left-wing view that all creatures are in some sense equal. The evolution of complex creatures is no mere epiphenomenon; it is one of the marvels of the universe.

Whether this argument leads us to a perfect Middle Way in the center of the left-right spectrum of views of life or whether it leads to what I have called a biased Middle Way (center-right) is of little consequence. In the former, diversification within a level of complexity, on the one hand, and ascents or descents in complexity, on the other, are seen as independent dimensions of the evolutionary process and given equal weight. That's fine. In the latter, changes in complexity, and especially ascents, are given more prominence. In this book, I adopt the center-right approach to correct an imbalance in evolutionary theory in which increases in complexity have not had their fair share of attention. If I succeed in persuading readers of this, I will probably age gracefully, shifting back gradually to a central Middle Way approach. If not, I will remain the biased "angry young man" of evolutionary theory, albeit well past my best-by date for the label, continuing to argue the case for the importance of the complexity issue.

So much for ideology. It's a subject whose influence on apparently objective matters should never be underestimated. But equally, it's dangerous territory for scientists, who are often accused of dealing with it naively by those, such as philosophers and sociologists of science, who are the relevant professionals. So I now head for safer ground—namely, the pragmatic rather than ideological reason for the comparative neglect of the complexity issue by past and present evolutionary theory.

One of the most famous American paleontologists of the mid-twentieth century, G. G. Simpson, once said: "Synthesis has become both more necessary and more difficult as evolutionary studies have become more diffuse and more specialized. Knowing more and more about less and less may mean that relationships are lost and the grand pattern and great processes of life are overlooked." This statement was then, and continues to be today, a remarkably astute comment on evolutionary biology. After the big idea of evolution was accepted in general, most evolutionary biologists concentrated on describing what kinds of evolutionary change occurred in particular lineages of creatures, in particular places, and over particular periods of time. Each piece of work of this kind is typically referred to as a case study.

Although there are exceptions, and we'll come to those toward the end of this chapter, most case studies are narrow in the sense that they focus on comparisons within a relatively small and tightly knit group of animals—often a group at the level of a family (for example, the dog family, the Canidae) or below. Evolutionary biologists have conducted this kind of narrowly focused case study not because of a perverse delight in evading the big picture, but rather because they understand that to achieve a reasonable degree of detail, and an in-depth knowledge of just what is happening in the group concerned, they need to focus on small groups if the task is to be manageable.

What I'm going to do now is take you on a rapid guided tour of five evolutionary case studies—three carried out on existing (extant) creatures, two on the fossil remains of extinct ones. Of the three on extant creatures, two involve large, externally visible characters of the creatures concerned, moth colors and bird beaks, while the other involves a microscopic internal character, a blood protein. All are at the famous end of the spectrum of case studies, which will allow me to cover them quickly and concentrate, in each case, on my main point—that the nature of such studies means that complexity issues get automatically pushed into the background, if indeed they feature at all.

Case study 1 involves the evolution of various species of moths, especially the species known as the peppered moth, from a starting point of having pale-colored wings to an end point where whole populations of the species concerned, usually those inhabiting polluted urban areas, are composed of individuals that have very darkly pigmented, or melanic, wings. (Actually, not only does evolution have no such thing as an end point, but in this case the process has recently gone into reverse, with melanics declining in frequency.)

This story is perhaps the most famous of all in revealing evolution in action on a relatively short timescale—of the order of a century or so. The traditional explanation is that rare melanic mutants, which showed up intermittently by chance in the original population of pales, became favored by natural selection in the days of the industrial revolution because they were less visible to bird predators than the pales when they were resting on soot-blackened tree trunks and branches. Previously, the pales had been fitter because their speckled pigmentation was well camouflaged against the background of natural, lichen-covered tree trunks.

This explanation has been challenged recently in a popular science book, *Of Moths and Men*, by the American author Judith Hooper, and some of the traditionalists have rebuffed her challenge. I'd guess that the nature of the selection going on in these moth populations may well be more complicated than was previously thought, but that's merely an interesting twist to an already-good story. There is certainly no question of the evolutionary changes concerned being caused by anything other than natural selection. They are not random changes, because they are too repeatable from place to place and from species to species for that to be a viable explanation.

Whatever precise way you choose to interpret this case study, it is clear that the newer form of moth, the melanic, is neither more nor less complex than the original pale form. This is emphatically an instance of diversification—becoming different—but remaining on the same rung of the ladder of complexity. The melanic form does not have more wings, or other body parts, than its pale counterpart. And although the chemistry of pigment production in the two forms is only imperfectly known, it does not appear that either form has a significantly more complex internal chemistry than the other. Again, in this respect, they are just different.

This finding should hardly come as a surprise. After all, we are dealing here with what is often described as microevolution: small changes that take place over a comparatively short timescale within the confines of a narrow group of creatures. The peppered-moth case does not even involve the origin of a new species, since the melanics and pales can interbreed quite happily. Evolutionary changes in complexity, in contrast, we might reasonably expect to take much longer than a few centuries.

Case study 2 involves the group of birds known as Darwin's finches. I probably don't have to tell you that these live on the Galápagos Islands off the west coast of South America and that they became famous from Darwin's study of them during and after his visit to the Galápagos archipelago while on his round-the-world trip on the *Beagle*, which was a major contributor to his ideas on evolution.

There are several species of these finches. Recent molecular studies have questioned previous views on the precise number, but from my perspective here this doesn't really matter. Whether there are two species or twenty, the story is, from a complexity viewpoint, the same.

The basic idea is that the original colonizers, which could have been just a couple of individual birds or a large flock, arrived on the islands from somewhere on the mainland. They multiplied and spread across all the different islands and islets and, over time, diversified—for example, in the sizes and shapes of their beaks. This represents adaptation to the local conditions, including the range of food items available, which differed somewhat between islands. The process of diversification sometimes produced a new species, but sometimes it merely produced an intraspecific variant form. As with the moth example, the precise nature of the selection going on has been debated. But again, from a complexity viewpoint, such debates are of little import.

If you go to the Galápagos and look at these finches, or if you consult an evolutionary biology textbook and look at glossy photos of them, you will see differences in the length, depth, and degree of curvature of the beaks; you will also notice differences in other characters, such as body size and plumage pigmentation—though in regard to the latter it should perhaps be noted that this group of finches is in general rather drab, in contrast to the very colorful species of finches, for example, the goldfinch, that can be found elsewhere.

Is a finch with a longer beak more complex than one with a shorter beak? Absolutely not. Recall our earlier working definition of complexity: the number of types of parts that a creature is made up of. Making a longer beak does not require more different types of specialist cells than making a shorter one; rather, it merely requires a slight adjustment in the timing or spacing of certain developmental processes. Again, as with the moths, this is a clear case of diversification within a level of complexity. No one has gone up or down the ladder; one rung has just become a little more populated with species than it was before.

Case study 3 involves the evolution of horses, and this study is largely in the paleontological realm. The very same paleontologist who drew our attention to the need to avoid losing sight of the big picture, G. G. Simpson, did much of the early work on this particular case study. Fossil evidence shows that over millions of years, horses have diversified in many ways. The characters involved are both large, such as overall body size, and small, such as the precise structure of the teeth. Let's focus on the former.

One thing that is clear from Simpson's (and others') studies of fossil horses is that the original horses—that is, those of the stem lineage—were quite small. In the context of today's mammalian fauna, they were more dog-sized than horse-sized. So one thing that has happened in horse evolution is an increase in body mass.

All of today's horse-group creatures, including zebras, are considerably larger than their distant ancestors.

Is a large horse more complex than a small one? Again, as with birds that have different beak lengths, emphatically not. Bigger horses have the same range of internal organs as small ones, and they have the same range of cell types. Admittedly there is an element of speculation here, because, while such assertions could be tested in living horses—for example, Shetland ponies versus racehorses—they cannot be tested in fossils. We cannot come up with an accurate figure for the number of cell types in stem lineage horses, because the details of individual cells are rarely evident in fossil material. (There are a few exceptions to this generalization, but they do not involve horses.)

We have, without really noticing it, moved from the realm of microevolution to that of macroevolution. That is, as we have moved from moths to finches to horses, we have shifted our focus in two ways. First, the timescale has lengthened—from centuries to millions of years. Second, we have extended our taxonomic scope from within species to between species, or from intra- to trans-specific, if you prefer. Yet despite this significant shift in focus, we are still not seeing changes in the level of complexity of the creatures concerned. This may be because such changes belong in yet another realm—that of megaevolution, a term coined by Simpson but relatively little used by subsequent authors (compared with micro- and macroevolution).

I have long thought that the underusage of Simpson's term "megaevolution" has been unhelpful in the development of evolutionary theory, and I will most certainly use it here. It refers to even longer-term and bigger-scale changes than "macroevolution." Although there is no clear line separating the two, neither is there one between "micro" and "macro," so this is not an argument for omitting the term. Indeed, if fuzzy boundaries and gray areas were adequate reasons not to invent terms for the realms on either side of them, much of biological terminology would disappear.

Case study 4 involves fossil mollusks from Lake Turkana in Africa. In the 1970s, two American paleontologists, Niles Eldredge and Stephen Jay Gould, came up with the theory of punctuated equilibrium. Their study of various lineages of animals led them to propose that evolution is not a slow, gradual, erratic process, as envisaged by Darwin and many others since, but rather a process in which nothing happens over long periods of time and evolutionary changes occur suddenly (geologically speaking) in short bursts that punctuate the long periods of stasis or equilibrium, thus giving the theory its name.

This "theory" generated much discussion, but most of this resulted, in my view, in the production of heat rather than light. Many of the views and counterviews were ideologically motivated (we're back there again), and sometimes the language used was designed to be inflammatory—as in "Puncturing Punctuationism," a chapter title in the popular science book *The Blind Watchmaker* by the English evolutionary biologist Richard Dawkins.

In my view, punctuated equilibrium is not a theory at all; it is a pattern. And it is indeed a pattern seen in many lineages of creatures in many parts of the world. But the interpretation of this pattern is another matter entirely. I am not convinced that the prevalence of such patterns argues against a standard Darwinian interpretation of the changes involved, as some have argued, though it remains a possibility.

Anyhow, one of the most famous case studies of fossil lineages arguing both for a punctuated pattern and for a non-Darwinian interpretation of the pattern was conducted in the 1980s by an English paleontologist, Peter Williamson, working at Harvard. This study, like so many of the best ones, was published in the leading scientific journal *Nature* (where Watson and Crick's paper on the structure of DNA had been published three decades earlier). It caused quite a stir at the time, but I'm going to ignore all that and instead ask my usual question.

Are mollusks that make shells of slightly different shapes from those of their ancestors—more tightly coiled, for example—any more or less complex than those ancestral forms? You already know that my answer is no. However, I have included this case study for a particular reason. It introduces the issue of what has sometimes been called the tempo of evolution (another term for which we are indebted to G. G. Simpson). The central question here, as we have just seen, is whether evolution proceeds gradually or in sudden leaps. Although I don't think that macroevolutionary leaps or punctuations are relevant to the evolution of complexity, an interesting possibility that we can address later is whether megaevolutionary leaps—sometimes called saltations—occur and, if so, whether they might be relevant to our central theme.

Case study 5 involves changes in the structure of the human blood protein hemoglobin. This protein is the main component of our red blood cells. Its function is to carry oxygen around our bodies and deliver it to all the parts that need it—that is, almost everywhere. It is a life-or-death molecule. If it works properly, we can live a normal life; if it is sufficiently mutant, which regrettably it is in some people, an early death is inevitable. But some forms of mutant hemoglobin fall in between these two extremes—they may cause problems, but also have some benefits. It is one of these, namely the one that causes the disease sickle-cell anemia, that I want to discuss.

As molecules go, hemoglobin is big. It consists of two identical halves, each of which has about three hundred of the protein building blocks we call amino acids. There are about twenty different amino acids, and each three-hundred-unit half of the hemoglobin molecule has a definite sequence of these units along its length. The mutant hemoglobin that causes sickle-cell anemia is identical to its normal equivalent in all but one of these. In other words, the degree of sequence similarity exceeds 99 percent. Yet the tiny difference of one amino acid out of so many can make the difference between life and death.

Notice that I said "can" rather than "does." The reason for this distinction is that the vast majority of animals, humans included, have two copies of most of their genes. Since proteins are made by genes, this means that if an individual has one copy of the gene that makes mutant sickle-cell hemoglobin and one that makes the normal version of the molecule, then that person will probably be fine. Another person, with both copies mutant, will be seriously affected and may die at an early age as a result.

The disease works something like this: if you have only mutant hemoglobin, the molecules fold up in the wrong way and distort the shape of the red blood cells that they populate; these cells become sickle-shaped and have seriously reduced oxygen-transporting ability. Given the severity of this problem, and the fact that some human populations have a reasonably high frequency of the mutant gene concerned, you might want to ask the following question: Why has natural selection not reduced the frequency of the mutant gene to zero?

In most human populations, natural selection has indeed done exactly that. The clue to why it has not done likewise in the others lies in their geographic locations. These populations are mostly in areas of central Africa where malaria is rife. By a strange coincidence, the effects of having one mutant sickle-cell gene include conferring resistance to malaria. Individuals with one mutant and one normal copy of the relevant gene—heterozygotes in genetic terminology—tend to have the best chances of survival because they don't suffer too much either from genetically induced anemia or from parasitically induced malaria. This effect creates a kind of balance of mortalities in the population, with normal homozygotes often dying from malaria, mutant homozygotes often dying from anemia, and heterozygotes not dying from either cause and so having a good chance of living a long life. In such situations, the beneficial side effects of an otherwise harmful gene cause it to remain at high frequency in the population indefinitely.

The kind of natural selection involved here is called balancing selection, in contrast to the directional selection responsible for the evolutionary changes we saw in earlier case studies—for example, from pale to melanic moths. In other respects, the two processes, in moths and men, are quite similar. Both involve natural selection working on single genes at a local population level. Both fall firmly into the category of microevolution.

And so to our usual question. Are humans with mutant sickle-cell hemoglobin any more or less complex than those with the normal version of the molecule? Again, the answer is no. If the wrong amino acid in sickle-cell sufferers had been an entirely novel one, outside the usual group of twenty, then I suppose it could be argued that these people had more complex hemoglobin than normal, because it would have more types of parts. But this is not the case. Rather, the wrong amino acid is a very common one; it's just that in this instance, it occurs in the wrong place.

The British biologist C. H. Waddington once said: "It is doubtful if anyone would have ever felt any need to resist the notion of evolution if all it implied was that the exact chemical constitution of hemoglobin gradually changed over the ages." I am particularly fond of this quotation. It warns against an overly reductionist view of life under which it is thought that a shift from organismic to molecular case studies will solve all the interesting questions of biology. Molecular studies, and especially molecular *genetic* studies, have, over the last half century, added hugely to our understanding of living creatures. To fail to acknowledge this would be a folly indeed. But these studies are not enough on their own. The big issues of biology, like the evolution of organismic complexity, require both molecular and higher-level studies, and, perhaps most important, the integration of the two.

We can now climb down from our fifth tree and ascend a path to a clearing on the hilltop in order to once again survey the whole wood. The name of the wood is, if you like, the Wood of Deafening Silence. The reasons for the silence—well, the main two anyhow—are now clear. Evolutionary theory has had little to say about changes in the complexity of creatures partly because of ideological issues and partly because most case studies of the evolutionary process have been of micro and macro, but not mega, scope.

This last point leads to both a hope and a problem. The hope is that new advances in understanding the evolution of complexity may arise from a very different kind of case study that has come to the fore in recent years. This kind of study involves megaevolutionary comparisons of important genes that, unexpectedly, have turned out to be shared across vast swaths of the living world—for example, from sponges to sperm whales. The problem is that if megaevolution produces changes in complexity while its micro/macro counterparts do not, then it must operate in a fundamentally different way from how they do. This conclusion runs against the current mainstream view that mega effects are simply the results of a multitude of micro effects accumulated over enormous lengths of time. Whether the mainstream view is correct remains to be seen.

CHAPTER FIVE

CORK PRISONS

This chapter marks the end of one journey and the beginning of another. The route we have taken so far has been of a philosophical and historical flavor. We have noted the special importance of ascents in complexity, but also the necessity to avoid thinking of all life-forms as arranged up and down a single ladder. We have tried to look at earthbound life from an external (Martian) perspective. And we have seen that, historically, the writings of most evolutionists have had little to say about increases in complexity. Now we begin to look at the story of these increases, and we note that they are of two kinds—embryological and evolutionary. This chapter deals with the basic building block for both of these complexity-enhancing processes: the cell.

WE all know that the human body is made up of organs. Some of these are visible from the outside—such as the eyes you are currently using to read this book. Others we know to be there, but we never see them, as they are internal. Hearts, brains, livers, and such are only seen by surgeons, soldiers, medical students, pathologists, and perhaps a few others. However, this does not stop us from believing in their existence. Indeed, we don't give it a second thought.

But what, in turn, are organs made of? This question can be answered in at least two ways. The less informative is that organs are made of different tissues. For example, a complete heart is mostly made of muscle tissue, but it also incorporates several other tissue types, including nervous and connective tissue. The more informative answer is that organs—and tissues—are made of cells.

At one level, cells are extraordinarily complex things. They consist of millions of molecules, of thousands of different types, interacting in countless ways at enormous speed, as we saw when watching the movie *The Story of a Cell* in Chapter 3. But at another level, we can ignore all this detail—fascinating to some, off-putting to others—because it is unnecessary for the task at hand. The goal I have set for myself here is to inspire you with thoughts about how both evolution and development produce creatures with many cells from a starting point of a single cell. To achieve this goal, I will have to share with you a few cellular details—like how some cells have a sort of surface Velcro that causes them to stick together, while some do not—but we can ignore many other details without ill effect.

Organs have been known about for much longer than cells. Since they are a feature of almost all animal bodies, not just human ones, they would have been discovered by prehistoric hunters. Cells are also a feature of all (well, nearly all) bodies. The delay in their discovery stemmed not from rarity but rather from smallness. Most cells cannot be seen without the aid of a microscope. Therefore, before we had microscopes, we were blind to the existence of these tiny building blocks. The first microscopes were devised around 1600 by a spectacles maker, Zacharias Jansen, and, like all inventions, began to evolve toward progressively better designs. This evolution led to such present-day wonders as the electron microscope. But many major discoveries did not have to await this evolution of design.

Cells were first seen, and given their name, by the English scientist Robert Hooke in 1665. He saw them in what might at first seem a rather unpromising material—cork. This material, although botanical, and so from the realm of the living, is effectively dead. What Hooke saw, in the thin slices of cork he looked at under his primitive microscope, were cross sections of regular sealed units a bit like the cells of monasteries or prisons, which is why he gave them their name. These units were defined by cell walls, rigid outer boundaries of plant cells (animal cells don't have them) that can survive the death of the cells they used to contain.

Animal cells are very like plant cells, despite this particular difference of the presence or absence of an outermost wall. All cells are bounded by an outer membrane. If a wall is present, then it is outside the membrane. Inside, cells typically have a central body called the nucleus that contains the genetic material. The rest of the cell—between central nucleus and outer membrane—is composed of a fluid material called cytoplasm (literally "cell material") in which many little bodies smaller than the nucleus can be found. These organelles, as they are called, are present in all animal, plant, and fungal cells, but are absent in bacteria, which have simpler cells lacking even a proper nucleus—but still bounded by an outer membrane.

It's one thing to see cells in specific source materials—whether a slice of cork or a slice of brain—but it's quite another to propose that all tissues and organs in all kinds of creatures are made up of cells. Yet that is just what two German biologists—Matthias Schleiden and Theodor Schwann—proposed in the 1830s. The cell theory, as it came to be called, was remarkably risky as scientific theories go, and yet turned out to be (almost) correct. Let me expand on both of these points.

Schleiden and Schwann, between them, examined many types of tissues from many kinds of creatures. But given the existence of millions of species, some of them with a very large number of tissue types, the *proportion* of all possible species/tissue combinations that our German heroes had examined was vanishingly small. Their claim that a cellular construction was widespread, even perhaps universal, in the living world was thus risky—because there were so many unknown cases in which another kind of construction might have prevailed. And of course, in the living world variation is the rule rather than the exception, in contrast to the physical world, where theories (like relativity) have a good chance of being true in a universal way—that is, with no exceptions at all.

Now, nearly two centuries and a multitude of case studies later, it is clear that the cell theory was about as right as any biological theory can be. It is not universally true, but it applies in more than 99 percent of cases. The exceptions are so few that they really do prove the rule, as the old saying goes. They come in two forms. First, viruses are noncellular, though whether they are truly alive (small creatures) or not (just big molecules) is a moot point. Second, some creatures, and some parts of other creatures, are made up of a thing called a syncytium. This term, like cytoplasm, is self-explanatory if you know enough Latin—which most of us, regrettably, don't (it was my worst subject at school, and I dropped it at the earliest opportunity). Syncytium, roughly translated, means "cells joined together."

There is a group of creatures that go by the delightfully downmarket name of slime molds. These are small squidgy things from the realm of strange organisms that are neither animals nor plants. One group of them has an adjectival prefix describing their special type of body construction: the syncytial slime molds. In this group, each individual creature has the following form: it consists of an outer membrane filled with cytoplasm (as usual), but it lacks the typical division into cells by lots of internal membranes; instead, many nuclei float adrift in an undivided cytoplasmic sea.

In some other creatures, although the vast majority of the body is divided up into cells in the normal way, one particular tissue (or occasionally more than one) is syncytial. For example, many parasitic worms have an outer layer of skin that is of this kind of construction.

So much for the exceptions. Although there are others that I have not mentioned, together they make up such a tiny proportion of the living world that we can ignore them. Most parts of most creatures are cellular—overwhelmingly so. Let's now take a whirlwind tour through the world of cells to get a feeling for their properties—what size they are, how long they live, how they communicate with each other, and so on.

Although all cells are small, some are much smaller than others. For example, a human egg cell has a diameter of about a tenth of a millimeter, whereas a human red blood cell is an order of magnitude smaller, at about a hundredth of a millimeter. Cells also vary considerably in shape. Egg cells are typically spherical, or in some cases ovoid, while red blood cells are shaped a bit like the wheel of a car (complete with well-inflated tire). Some cells that are tightly packed together are approximately cuboid, each having six juxtaposed neighbors. Others are elongate, or have long projections, as in the case of the nerve cells that snake through our bodies conveying messages over long distances.

Cells vary a lot in time as well as space. The typical life span of a cell can be just a few hours or days, as is the case with the cells that line our stomachs. The brevity of their existence is unsurprising, given that they are constantly bathed in hydrochloric acid. Other cells live for weeks or months, an example being those wheel-shaped red blood cells that circulate around our bodies for that sort of time, before they are destroyed and replaced by new ones. Plasma is a kinder medium to live in than an acid. Yet other cells live for years. This applies to many of our brain cells, which, when they eventually do die, are not replaced—lucky, then, that we have a larger-than-necessary number to begin with.

While the wide variety of cells is impressive, what is more impressive still, and more relevant to my cause here, is the uniformity that lies beneath the variety—that is, the structures and processes that most or all animal and plant cells have, regardless of whether they are big or small, fat or thin, short- or long-lived. I have already mentioned these *structures* (nucleus, organelles, and so on). What I now want to focus on are the *processes*. These are crucial because of course cells are not static entities; rather, they are hugely dynamic, constantly in a state of flux. This flux is the business of life.

There are several different types of flux, and I am going to concentrate more on some than others. The reason for my selectivity is that we will be moving on, in the next chapter, to look at how cells manage to form embryos, and, ultimately, adults. So the most relevant cellular processes are the ones that enable—even drive—this incredible developmental journey. The least relevant are what are sometimes called the housekeeping processes of cells—those that keep the cell alive by taking in nutrients and metabolizing these substances to provide an ongoing source of energy. I'm going to take all these cellular housekeeping processes for granted.

The three most important things that cells do beyond simply staying alive (or dying, sometimes) are dividing, differentiating (becoming different), and moving. Let's take them in that order. In each case, I will concentrate on aspects of the process concerned that are most relevant to embryonic, and subsequent, development.

A typical cell division that happens, for example, in the growing biceps muscle of a month-old chimpanzee involves two main things: division of the genetic material; and division of the cell itself. Both aspects of the overall process are, as you might imagine, hugely complicated at the molecular level. Whole books have been written about them. But luckily, we can evade most of the detail. The most important thing to know is that prior to the split of the parent cell into two daughter cells, the genetic material is duplicated in its entirety, and each of the duplicate sets of genes is eventually enveloped by a new nuclear membrane, thus retaining its integrity, within what becomes one of the daughter cells.

This duplication is no mean feat. Thanks to the Human Genome Project, we know that every individual of our own species (and probably the chimp, too) has about thirty-five thousand different genes. Each of these genes consists of a stretch of DNA that is thousands of units (nucleotides) long. Precise replication on this grand scale of numbers condensed into a microscale of space is almost incredible. Yet it happens all the time. It is happening in my body as I write and in yours as you read. It is not confined to embryos, because most adult cells that die need to be replaced, and the only way to do that is by division of those that remain alive.

A corollary of the precise duplication of the genetic material with each cell division is that all cells have a full complement of genes. Since it is the genes that make the main players in the chemistry of a cell, namely proteins—both enzymes and others—there is a puzzle here. If all cells have the same genes, how do they become so very different from each other? How, for example, do some cells become muscle cells and others nerve cells? In other words, how do cells differentiate, to use the technical but self-explanatory term?

The answer to this question is that although almost all cells have a full complement of genes, different genes are switched on (and thus making proteins) in different cases. For example, the genes that make the contractile proteins that belong in muscle cells are switched on in those cells but switched off in others, such as brain cells. At an early stage in our understanding of these things, this was known as the variable gene activity theory of cell differentiation. Like the cell theory, the variable gene activity theory has stood the test of time, and is known to be generally true (say, again in more than 99 percent of cases), but with the occasional exception that living systems always seem determined to provide.

From a starting point of a single cell, it would in theory be possible, given a supply of energy, to produce a large multicellular creature using only the two processes described above—division and differentiation. As cells multiplied and the embryo grew, all that would be necessary would be for cells to differentiate into the right types at the right times and in the right places. Although some stages of development do indeed work mainly in this way, others are characterized also by cell movements—in some cases large-scale streaming of whole populations of cells from one part of the embryo to another.

Movement more obviously requires an explanation than lack of movement. However, it makes sense to deal with the two together. Whether a cell stays still within a block of tissue, like a muscle, or whether it "decides" to head off on a long trek past many of what were its neighbors has a lot to do with its outer membrane, and in particular with certain types of protein molecule that may (or may not) be peppered over that membrane.

I referred earlier to a sort of intercellular Velcro that makes cells stick together. There are actually several different types of this Velcro (whose molecular nature we needn't delve into). Cells with one particular type tend to stick together with other cells that carry the same type. That's how blocks of tissue manage to stay together rather than fall apart. But given that fact, it becomes clear that the only cells that will be free to move will be those that have not yet acquired their Velcro, or those that had it and then, perhaps temporarily, lost it.

Lack of this Velcro is, if you like, a passive factor. If you are a mathematician, you might say that this lack is a necessary but not sufficient condition for a cell to move. If you are not a mathematician (statistically more likely, I suppose), you might, alternatively, say that the absence of a force for stasis is not the same as the presence of a force for motion. And, whichever way you chose to put it, you'd be right.

So what is it that makes cells move, rather than merely allows them the option of moving? As with so many of the questions I have been posing in this chapter, this one can be answered in a general or technical way, and, as with the others, I am going to opt for the former kind of answer.

Cells move because they receive a signal of some sort. There are lots of different ways that cells can communicate with each other. Some intercellular signals move only from one cell to its immediate neighbors; others are capable of traveling across many cells but are still restricted to a tiny part of the embryo (or, more generally, the organism); and others can be transported right from head to tail—for example, in the case of hormones that suffuse the body and have developmental effects all over the place.

Here's an example of a typical cell-signaling event, if there is such a thing. One cell secretes a specific protein molecule. This molecule travels through the labyrinthine network of cluttered corridors between cells called the extracellular matrix, and eventually is grabbed by a specific receptor molecule that is sticking out through the outer membrane of what you might call a target cell—that is, one that the cell we started with "wants" to influence in a developmental way. The secreted protein is too big to get through the membrane of the target cell, but that doesn't matter, because the other end of the receptor molecule sticks out, too—in this case into the cytoplasm. When the outside bit of the receptor picks up the signal molecule that has arrived from afar, the whole receptor—a transmembrane molecule, as you can now see—does a sort of wiggle that transmits to the cell's interior the information that the signal has arrived. This information is then passed from molecule to molecule within the cytoplasm in what is best thought of as a microscopic relay race (but maybe with only one team). The final runner sprints through one of the convenient holes in the nuclear membrane, attaches itself to a particular gene, and, with a little help from some other friendly molecules, switches the gene on or off.

What happens next depends on what the gene that has just been switched on or off does—to put it another way, what kind of protein it makes. We started looking at signaling and communication in the context of how you could make a cell move. Well, some genes make proteins that go out to the surface of the cell and, instead of staying in one place, like the Velcro proteins, get recycled in a way that makes the cell move past its neighbors as if it had multidimensional caterpillar tracks.

But intercellular communication isn't important only for movement; it's important for the other two of our "big three" cell processes, too—namely, division and differentiation. The gene in the target cell that gets switched on by an incoming signal may not make a caterpillar-track protein. It may instead make a cell-type-specific protein, like hemoglobin, which we encountered earlier. In that case, cell differentiation rather than movement is the result. And in other cases, the gene that gets switched on may make a protein that initiates replication of the genetic material—this time, cell division will be the outcome of receipt of the initial signal.

So the three crucial things that cells do (in addition to their standard internal housekeeping functions), namely division, differentiation, and movement, are all driven by signals that arrive from other cells. And in any particular case, the cell that responds to a signal may, among other things, send out a signal of its own. The embryo, and the adult into which it gradually transmutes, are highly integrated communities of cells all talking to each other in many languages simultaneously at a speed that makes the stereotypical native French speaker, whose words all rush out together in a torrent, sound slow.

This chapter has been centered on one of the main players in the developmental game, and indeed the game of life more generally: the cell. But in delving into some of the things that cells do, we have encountered some of the other main players, too—in particular, genes and their protein products. In the next two chapters, I'll focus on the business of how to build a body, taking first a cellular view and then a genetic one. Since these discussions will be biased in the corresponding ways, I should end the present chapter with another warning about philosophical baggage.

There is a book sitting on the bookshelves behind me called *The Genetic Basis of Development*. Many authors write of the genetic *control* of development. Other authors rebel against what they see as genetic imperialism and downplay the role of the genes—in some cases far too much. We need to avoid both overly genocentric and overly genophobic stances. In the end, any creature develops into whatever is its adult form from a starting point of an egg because of a balanced, two-way process in which genes and cells tell each other what to do. Even that description, inasmuch as it suggests only a two-way reciprocity, is a mere caricature of reality. The complexity of the communications that go on among genes, proteins, other molecules, and the cells in which most of these players are contained, and yet which are players in their own right, too, is almost beyond our imagination. The rapid multilingual conversation is probably the best of the many metaphors I have used so far, yet even that doesn't really do it justice.

So now, with as little baggage as possible, we ascend from the realm of the cell to that of the whole creature, whatever it may be. And as we shall see, despite the vast diversity of creatures that can be found in our biosphere, there are, paradoxically, many similarities in the way they are made.

CHAPTER SIX

BUILDING A CASTLE OF CELLS

We saw the parallel between sandcastles made out of sand grains and creatures made out of cells in Chapter 3. We also saw the lack of a parallel in that sandcastles cannot make themselves whereas creatures can. But how does this "self-making" work? How do fertilized eggs become embryos and embryos become adults? This developmental rise in complexity is the focus of the present chapter.

WHEN I was a student, some thirty years ago, it was often said that development—that magnificent journey from egg to adult, which most creatures embark upon, even though only a minority reach their destination—was the biggest unsolved problem in the whole of biology. And in my view, that was indeed true at the time. In terms of pure, as opposed to applied, biology, there were the cell theory, the theory of natural selection, and Mendel's laws of inheritance, which had recently been given a molecular dimension by Watson and Crick. So there were grand general theories in the areas of what organisms are made of, how they pass on their genes to the next generation, and how, in the longer term, evolution could change both organisms and their genes.

Contrast these remarkable achievements with the state of developmental biology in the same period—the early 1970s. There was no general theory of development. Some biologists were even arguing that there could never be such a theory, because development was a plethora of different subprocesses, making it too bitty to be explained by any grand, sweeping generalizations.

At that time, developmental biology, then more often called embryology, had three main strands—descriptive, comparative, and experimental (respectively: describing development in a particular creature; comparing the development of different creatures; and deliberately perturbing development to see what happens). The first strand dates back at least as far as the descriptive studies on chick development made by the Italian scientist Hieronymus Fabricius in the late sixteenth and early seventeenth century, and published (posthumously) in 1621. The second can be traced back to those early-nineteenth-century German nature philosophers we encountered in Chapter 1. The third was started around 1900 by another German, the pioneering experimental embryologist Wilhelm Roux. In all three cases, there were probably earlier studies—perhaps *much* earlier—than those mentioned. For example, Aristotle gave some thought to development when he classified animals into eight groups sometime around 350 B.C.

As we can see now with the benefit of hindsight, these three strands were not enough. There was a fourth waiting in the wings, without which a general theory of development could not be formulated. This genetic strand did not come into its own until 1984, with the discovery of a gene sequence called the homeobox that forms the core of Chapter 13. But, as with the other strands, pinning down an exact origin is very difficult. Even in 1970, a few pioneering souls were doing great work in what would become, at least in my view, the most crucial strand of developmental biology, namely developmental genetics. Perhaps the most important of these was the American geneticist Ed Lewis, who was already publishing interesting work in the 1960s on how mutant genes affected the development of flies, and how we might learn from such effects what the normal versions of the genes concerned did to send development in the right direction.

We'll look at what has emerged from this genetic strand, in terms of a general theory of development, in the next chapter. But the horse needs to come before the proverbial cart. In this case, the facts of development are the horse, while the cart is that elusive general theory about how development happens.

Straightaway, we run into a problem. The facts of development are very different from creature to creature. Even at the start of embryogenesis, when the fertilized egg begins to divide and multiply, this process happens in different ways in different animals. Later in development, the events that occur are even more different, to the extent that some animals will develop eyes while others will not, and so on. How *similar* developmental processes—the ones we'll look at in the next chapter—produce wildly *different* outcomes in different animal groups is a fascinating, but ultimately soluble, paradox.

How do I proceed with the story of developmental facts? An encyclopedic listing of these facts in species after species would send us both to sleep long before we got to the end of our writing/reading. Instead, it makes more sense to be selective, and to mention just a few kinds of animals as representatives of what happens in larger groups. This approach reflects the traditional way of doing things in professional work in this area. It is not unusual, for example, to find a description of the development of "the frog," notwithstanding the fact that there are many different species of frogs, some of which develop to adulthood without going through a tadpole stage on the way. And some of the processes discovered in research on a frog may even be used as examples of how things happen in vertebrates in general. For example, the ways in which some cells differentiate into muscle or nerve cell types in a frog and a human have much in common.

So my first strategy is to be selective. My second strategy, in this chapter, will be to start at the beginning—the egg—and to head for the adult; it would hardly make sense to adopt any other sequence. I do not wish to follow moviemakers into the realm of flashbacks. Strategy 3 will be my generic one of being as brief and undetailed as possible. And finally, strategy 4 will be to inject a little (temporary) ideology before we come to grips with any facts at all.

In 1999, four colleagues and I started a new scientific journal titled *Evolution and Development*. We invited the London-based media don Lewis Wolpert to write an opinion piece for our inaugural issue. In it, he made the following statement: "The cell is evolution's most magnificent achievement and embryonic development is merely a baroque elaboration." Should we believe him? The short answer is no; but here is a slightly longer one. Personally, I go along with Wolpert on this issue only up to a very limited point. He's right in one particular respect. The cell certainly is a magnificent evolutionary achievement. But so is embryogenesis. The idea that getting from egg to embryo (or, later, to juvenile and adult) can be described as "merely" anything, and especially as merely "baroque," is manifest nonsense. Wolpert was being deliberately provocative when he made this statement, just as Stephen Jay Gould was being deliberately provocative when he described humans and other complex creatures as a mere epiphenomenon in the history of life.

In the following, then, not only will I be telling you what happens in some parts of the developmental process in some creatures, but I will also, in doing so, be backing up my point that this process, just as much as the cells on which it is based, is one of nature's foremost wonders and not something that can be written off as an irrelevant elaboration of Wolpert's supposedly all-important cell.

Without further ado, let's get down to the business of how castles of cells—or organisms, if you want to use a more conventional descriptor—get built. And, as I said earlier, we'll start with that crucial first cell, the fertilized egg (or its unfertilized equivalent in cases of asexual reproduction). The first process to occur in the development of almost all animals is a thing called cleavage. In the old days of giving handwritten manuscripts to secretaries who dutifully turned them into type, I often found that I had to correct "clearage" to "cleavage" in the draft typescript, I think because the secretary concerned felt it unlikely that I really meant to use the word "cleavage" in a scientific article.

There is, however, a good reason for the use of this particular word to describe the first cell divisions that happen in the developing embryo. In most cases, cleavage is not accompanied by overall growth, so what is happening is, literally, the cleaving of one big cell, the egg, into lots of little cells. Exactly how many "lots" is depends on the creature concerned, but, to an approximation, it will suffice to think of the end point of cleavage as a ball of a hundred cells. Note, of course, that the idea of discrete stages of development, each with a so-called end point, is inaccurate. It is merely a device we biologists use to make the task of understanding and explaining a very long and complex process a bit easier than it would otherwise be.

Cleavage can occur in different patterns. Two important patterns, each characterizing large groups of animals, are called spiral and radial. These terms describe the ways in which cells arrange themselves, with newly formed groups of cells either being twisted around, clockwise or anticlockwise, relative to their progenitor cells (spiral) or lying directly on top of their progenitors (radial). Also, the ball of cells that cleavage, whether spiral or radial, produces varies somewhat between animals in its precise form. As well as varying in the total number of cells, it varies in how different the cells are (generally not very different), and in whether the ball of cells is solid or hollow. In cases of the latter form, which predominate, there is a fluid-filled cavity in the middle. Bear in mind that even these minor differences between animals at the earliest stage of development are a product of evolution.

The second process of development, which picks up where cleavage leaves off, is called gastrulation, because it leads to a stage called the gastrula. If you don't like technical terms, call it "poking in and rearranging"; why this is an appropriate phrase will soon become apparent.

At this point, we'll temporarily think of the hollow ball of cells that is the most usual result of cleavage as a balloon. Of course, the relative volume of the internal cavity and the thickness of the layer of tissue surrounding it are very different in the embryo and the balloon, but let's ignore that for now.

This case is the first of many in which we will see that nature is often illogical, or even perverse, in that it does not work in the same way as an engineer works when designing and building a machine. Since most adult animals have an internal cavity—the gut—it would seem logical to make this structure directly out of the initial cavity that forms in the middle of the embryo. But this is not what happens. Rather, the original cavity shrinks and disappears, and a new one begins to form. Using our balloon analogy, we can picture it as someone poking a finger into the balloon at a particular point, causing part of the outer tissue to form a projection into the internal cavity. In real embryos, outer-layer cells begin to move into the interior in a similar way. Eventually, the inwardly protruding tube that they form breaks through at the other side of the embryo, so we have a complete passageway from what will become the anterior end of the embryo to what will become its posterior. It is this tube that becomes the gut.

Now we encounter another apparent illogicality. In some animals, the first-formed end of the passage will become the mouth, while in others, including ourselves, it will become the anus. In fact, this fundamental difference is a basis for recognizing the two main groups within the animal kingdom, whose scientific names—protostomes and deuterostomes—translate as mouth-first and mouth-second. (I suppose that the alternative label of anus-first for our own group was considered too unsavory.)

Apparent illogicalities in the realm of life are the result of an evolutionary, rather than engineering, origin. One of the famous facts about later development in humans is that, at a certain embryonic stage, we have structures that are like rudimentary gill slits. If humans were built in factories, like the androids of science fiction, there would be no such stage in the fabrication process. But since instead we have evolved through a long line of ancestors, some of which were fish, the persistence of transient embryonic gill slits is entirely understandable.

Anyhow, back to gastrulation. I described it earlier as "poking in and rearranging." We have now seen why this is a reasonable description. At the start of the process, the embryo is roughly spherical in most cases, whereas at the finish it is elongate, with head and tail ends, and a primitive gut running between them. Although gastrulation involves cell division and (limited) cell differentiation, its most characteristic feature, in contrast to cleavage, is cell movement. Clearly, this stage of embryogenesis involves much change in the amount and type of the intercellular Velcro that we encountered earlier. During the period of maximal streaming of cells, there must be precious little Velcro in place around the periphery of the cells concerned, compared with the situation that applies in later stages characterized more by differentiation than by movement.

So in the first stage of development, cleavage, the most conspicuous process is cell division, and in the second, gastrulation, it is cell movement. In the third stage, with the pleasantly self-explanatory name of organogenesis, the most conspicuous process changes again. This time it is cell differentiation.

It's time to pause for a cautionary thought. Organogenesis means many things to many creatures. In an eyeless subterranean centipede, the organs generated include such things as a primitive brain and heart that are in common with a surface-dwelling centipede cousin. But eyes are clearly not generated in the first case, whereas in the second, at least two eyes, and sometimes more than twenty, are produced. Equally, in the world of vertebrate animals, some, such as ourselves, generate limbs; others, such as snakes, do not—though some form little limb buds in memory, as it were, of their four-legged ancestors.

Even though different creatures make different arrays of organs, certain fundamental processes are always involved, whatever organ is being made. Recall that organs are made of tissues and that tissues are made of cells. Hence the centrality of cell differentiation in organogenesis. If you want to make a heart, you need to send lots of cells down a route of differentiation that specializes them as muscle cells; whereas if you want to make a brain, you do not want this particular route—instead, you want to make nerve cells.

Since cell differentiation involves switching some genes on and others off, it makes sense to leave further discussion of it to the following chapter. But we're not yet ready for genetic theories of development, because we haven't yet quite finished with developmental facts, despite organogenesis often being regarded as the final stage of development. Even sexual maturation, which is typically the last significant developmental process to occur, can be thought of as part of the organogenetic phase—which means, incidentally, that our three-way division of development into cleavage, gastrulation, and organogenesis is a very asymmetrical one.

I have yet to mention something very important, namely a thing called either pattern formation or morphogenesis. (I use these terms interchangeably; some authors consider the two to be subtly different, but for reasons that do not persuade me to believe them.) A good way to approach this issue is to think of different muscles in your own body. Perhaps our most well-known muscle is the biceps of the upper arm, made famous by Popeye because his bulged to incredible size (and at incredible speed) on consumption of a can of spinach. Real human biceps are a weedy equivalent. They are less bulky, but of the same general shape—it is often described as a spindle shape.

Contrast this with the muscles of the abdominal wall (the abs) and with those that can produce horizontal ridges on our foreheads—for example, when we are asking questions. Both of these groups of muscles involve the tissue being formed into something more akin to a sheet than a spindle. The difference between a biceps muscle and an abdominal wall muscle is not of the same kind as the difference between a muscle and a brain. Exactly the same range of cell types is involved in both muscles. So the explanation for their different appearances is not to be found in the realm of cell differentiation—that is, the mostly internal changes through which an initially generalized cell becomes some sort of specialist cell.

It now becomes clear that a higher-order process is going on alongside cell differentiation. This process of pattern formation involves whole populations of cells. It must be based to a large extent on intercellular communication. Such a basis also applies to the process of differentiation: cells specialize in a particular way because they receive signals telling them to do so. But in pattern formation, both the nature of the signals and, even more so, the response to their receipt are different. This time the response is to divide in one particular plane or another at a particular rate, or to stop doing so. Movement of cells may also be altered in response to pattern-forming signals.

We'll return to this important topic in the next chapter and look at it in a more explicitly genetic light. For now, just a few developmental facts remain, all in the realm of late development—for example, growth.

Let's take a human perspective here and think of our own development. You might want to insist that there are really *four* stages of development—the three I dealt with earlier, plus growth. After all, most organogenesis of a human occurs in the embryo (and indeed within its first couple of months). Subsequently, most development, from two months to almost twenty years, consists of the growth of already-formed organs. It's not quite as clean a separation of stages as that, since things like breasts and wisdom teeth form rather late in the growth process. But, such complications aside, we could say that human development has four stages and that the fourth is characterized by the same basic cell process, namely division, as the first.

Now let's take a non-humanocentric view and think about butterflies. Here, in one of the pinnacles of the insect world, we find a fifth developmental process, metamorphosis. A butterfly, like a human, starts off life as a fertilized egg. Also like a human, it undergoes cleavage, gastrulation, and organogenesis while it is an embryo—albeit enclosed in an egg case rather than a womb. And, especially after hatching, it grows. A tiny caterpillar becomes a middling-sized caterpillar, which in turn becomes a large caterpillar, though these terms need to be interpreted differently in absolute size measurements depending on which kind of butterfly we are considering.

But there is no such thing as an adult caterpillar. Continued growth is not enough to complete the life cycle. At some point, the large caterpillar must stop eating, stop moving, harden its skin into a chrysalis, and, hidden away inside this temporary fortress, reconstruct itself into a fundamentally different type of living machine. The same cellular processes that characterize embryogenesis recur to drive metamorphosis. Again, cells are dividing, moving, and differentiating. But there is another cellular process that comes to the fore here, and which, although I haven't yet mentioned it, also plays a role in the embryo: cell death.

In metamorphosis, many, and probably most, of the caterpillar's cells are broken down and replaced with different ones. In other words, cell death is obvious and widespread. In embryos, its role is usually more subtle. An interesting example occurs in the development of vertebrate feet. We humans do not have webbed feet, because cells in the interdigital regions of human embryos undergo programmed cell death. In duck embryos, interdigital cell death is much reduced.

In a large, complex animal, such as a dolphin or a human, there are trillions of cells, each belonging to one of hundreds of cell types. How many ways are there to arrange these cells? You don't need the services of a mathematician to tell you that the answer to this question is an almost infinite number. Given that fact, the limited variation in developmental outcome within each species is almost miraculous. People vary in height, weight, shape, skin color, eye color, and personality, to mention just a few of our many traits. But the variation is minuscule given the number of *a priori* possibilities. Even a human with the right number of cells of every type could in theory take the form of a cube, with the cells of each type forming one of the layers of which the cube is made up. Of course, even if such an extreme mutant pattern formation were possible (it isn't), it would be lethal at an early stage in development because the "human" concerned would not function. If, for example, all blood vessel cells constituted one layer rather than a treelike system that pervades the whole body, most cells, and hence the developing organism, would die.

Some extreme variations of development *can* be produced as the result of mutation, as we'll see in the next chapter. The reason that they don't persist is just as in the hypothetical case above—genes causing extreme disturbances to the developmental process are removed from populations by Darwinian natural selection. Yet despite this, most species as found in nature rather than in a geneticist's laboratory exhibit not only slight quantitative variation among individuals in characters like size and shape; they also exhibit qualitatively different outcomes of the developmental process.

The most obvious of these is the difference between males and females, which is often pronounced. It's certainly quite impressive in humans, and it is even more so in some other animals, where the sexes can differ enormously in size, pigmentation (think of the contrast between a peacock and a peahen), and other features.

As well as different developmental outcomes in the two sexes, we ought to note such phenomena as winged and wingless forms of certain insects, as in many species of greenflies; the social castes of ants and bees, such as queens and workers; and, in the plant kingdom, the ability of some partially submerged water plants to produce completely different leaf forms above and below water. These different outcomes are often the result of signals from the environment being picked up by the developmental system and used to turn certain genes on and off in particular cells or groups of cells, just as internal signals (from one cell to another) can also bring about gene switching, as we have already discussed in outline, and will consider further in the next chapter.

CHAPTER SEVEN

DANCES WITH GENES

We live in the Age of the Gene. Recently, the entire complement of human genetic material (our genome) has been documented, and we know that each of us possesses about thirty-five thousand genes in total. Only a subset of these is directly involved in controlling the developmental process, but this subset is of huge importance. Here we look at development from the viewpoint of how developmental genes interact with one another and with the cells in which they are embedded, in a dance that has similar features across the entire animal kingdom and yet, paradoxically, produces very different outcomes in different animals.

THE developmental process is truly amazing. It produces increases in complexity in a tiny fraction of the time that evolution took to produce similar increases. From our last single-celled ancestor to the first protohuman took at least 600 million years. Human development produces an adult from a single-celled starting point in less than 20. In other words, development works about 30 million times faster than evolution. If you use a more rapidly developing creature, such as a small fly that reaches its million-cell-plus adult form in a mere ten days, then the ratio of developmental and evolutionary paces becomes even greater.

In the previous chapter, we looked at some of the actual happenings in the development of particular creatures, using the cell as our focal point. So we saw that cells divided, moved, differentiated, and died. We also saw that at the level of populations of many cells, there was an additional phenomenon of pattern formation, wherein groups of cells adopted particular shapes. We noted, at many points in the story of what happened in development, that the switching on and off of genes was a key part of the process. It's now time to investigate this phenomenon further by shifting our main focus from cells to genes.

Ideology is always just around the corner in biology's big issues. And often, as now, it's worth reaching around that corner and dragging it into view for a while before proceeding. Recall my earlier mention of genetic imperialism, the much-championed yet also much-criticized view (with its title provided by the critics!) that genes are the key players and in some sense control the developmental process. There is also a counterview that genes are merely responders to signals they receive and so in a sense have a more passive role, with the really active players in the developmental game being proteins (themselves, of course, made by genes, which takes us round in a circle). Even some predominantly genetic books express this view. For example, in the prologue of his recent book on gene regulation, the British geneticist Bryan Turner says that in general genes "do as they are told."

So who is in control? Is it the genes, the proteins they make, the cells in which genes and proteins interact, or the signals traveling between cells (often particular types of proteins themselves) that in some cases derive ultimately from some environmental cue—as when the sex of a turtle is determined by the temperature of its surroundings?

The most sensible answer to these questions, in my view, is that control works in various directions, so there are many kinds of key players, not just one. In fact, the very word "control" may be unhelpful. I prefer to talk of interaction rather than control. If we focus on a particular cell, signals coming from afar interact with components of the cell's outer membrane, which in turn interact with other molecules inside the cell, some of which move into the nucleus and interact with genes. A gene that is switched on as the end result of this process is most readily seen as a responder. But of course there is no such thing as an end result in such matters. The gene that gets switched on may make a protein that gets secreted from the cell and picked up by another, where it acts as a signal and starts the whole process over again. Looked at in this way, the gene is more readily seen as an initiator than as a responder.

Both ways of looking at developmental processes are equally valid—they differ really because they adopt different (and arbitrary) starting points in a continuous flux of complicated events. This point reinforces the view that it is better to think in terms of interaction than control. Genes, proteins, other molecules, and the cells in and around which they are found are all taking part in a sort of dance, the final outcome of which, at least in the luckiest of creatures, is the production of an adult. Even calling that outcome final is of debatable accuracy, as highlighted by the clichéd question of what came first, the chicken or the egg.

So far, all I have done in this chapter is reemphasize the point I made at the outset that we should travel with as little philosophical baggage as possible. It's always useful to rub in important points. But equally, it's inadvisable to belabor them, as I would be doing if I persisted further in an ideological vein. So now it's time to get down to business and look at (*a*) some examples of dances with genes and (*b*) some general theories of development that have arisen from genetic approaches to understanding developmental processes.

I'm going to concentrate on two processes at the cellular level:

cell differentiation and pattern formation. I'll take them in that order.

How do cells become different? For example, how does one cell become a muscle cell, another a skin cell, yet another a nerve cell? Back in the 1950s, when the nature of genes had been established, and the way they make proteins worked out at least in outline, there were two main possibilities. Since different cell types were characterized by different predominant proteins, it seemed logical that the process of differentiating involved either the loss of different genes in different cell types or the switching on and off of different genes in different cell types. In the first case, each cell would have only a smallish subset of the whole genome. In the second case, all cell types would contain, within their nuclei, a complete genome copy, but with different genes switched on and off in muscle cells, skin cells, nerve cells, and so on.

As we saw in the last chapter, it is now known that, with only a very few exceptions and complications, the latter process, differential gene switching, is what happens. One exception that I have not yet mentioned is the human (and more generally the mammal) red blood cell. These cells do not contain, within their nuclei, a copy of the complete genome, as stated above. In fact, they lack any genes and do not even have a nucleus—a rarity indeed among the cell types of animals generally. They are, however, packed with a particular cell-type-specific protein—hemoglobin, as we saw earlier when examining the inherited disease known as sickle-cell anemia.

How does a cell with no genes, and thus no ability to make proteins, come to be full to the brim with one particular protein? The answer lies in the cell lineage from which the mature red blood cell arises—in its ancestral cells, if you like. (We can easily use evolutionary terms to describe development; indeed it has become commonplace to do so.)

Earlier in this cell lineage, differentiation involves the switching on of the genes that make hemoglobin, with rapid and sustained production of this particular protein. Later, part of the differentiation process in this unusual case involves the loss of the nucleus. That doesn't matter, as the mature cell does not need to make any more hemoglobin. Rather, it is free to meander around the body, carried along with the flow of the blood, doing its crucial job of delivering oxygen to the various organs, until eventually, after perhaps a few weeks, it dies and is replaced by another whose origin is the same as its own—a progenitor cell that *does* have a nucleus.

So we have a general theory of cell differentiation. We know that almost all cell types possess a complete genome, and that the genes not needed in any particular cell type are simply switched off—they are not lost or destroyed. One beautiful proof of this was provided by experiments conducted in the 1960s by the British biologist John Gurdon, who combined the nucleus from a differentiated cell late in frog development with the cytoplasm of an egg cell whose nucleus had been deliberately destroyed, and discovered that the signals present in the egg cytoplasm could reactivate genes that had been switched off. Indeed, in a few cases, these hybrid cells went on to produce advanced developmental stages such as tadpoles. These experiments preceded the production of Dolly, the famous cloned sheep, by some thirty years, a fact that went unnoticed by many journalists of the 1990s who wrote inaccurate articles describing Dolly as the first cloned animal.

Let's stick with eggs, but shift from frogs and sheep to flies. One particular fly—a fruit fly—has been a favorite workhorse of geneticists since the early twentieth century, and, about half a century later, began to be used for *developmental* genetics, a subject that since then has mushroomed into a major industry, one that is producing some of the most exciting discoveries in the whole of biology.

The fruit-fly egg exhibits a gradient in a protein called bicoid. This strange name translates as "two tails" and describes the appearance of abnormal embryos in which it is absent. In normal embryos, the bicoid protein is at high concentration at the anterior end of the egg, and at progressively lower concentrations as we go from anterior to posterior. Because some genes are switched on by high concentrations of this protein, others by low, the concentration gradient conveys "positional information" that is used to begin the patterning of the embryo. Ironically, this very useful developmental phrase was invented by Lewis Wolpert, who, as we saw in the previous chapter, mischievously wrote off development as "merely a baroque elaboration" of the cell.

I said earlier that general theories of development were impossible at a previous stage in the history of developmental biology when it only had three of its current four strands—descriptive, comparative, and experimental, but not genetic. That, like so many statements about the history of science, is only true up to a point. In fact, an American developmental biologist, C. M. Child, came up with the theory that concentration gradients could be major players in developmental processes way back in the 1930s. In this case, the genetic strand did not initiate the theory, but it did flesh it out and give it enough substance that biologists generally would believe it.

Now, having stuck with eggs and changed animals, let's stick with the same animal and fast-forward to a later developmental stage. We'll consider the development of segments in fruit flies, and one aspect of their development in particular—their "identity." In any segmented creature, the most numerous group of such being the million-strong insect group to which the fruit fly belongs, it is useful to distinguish between (*a*) developmental processes that determine the *number* of segments and (*b*) the somewhat different processes, involving different groups of genes, that determine the precise structure of each segment, or its *identity*—for example, whether it is a thoracic or abdominal segment.

Segment identity is to a large extent an issue of pattern formation rather than cell differentiation. Consider, for example, two particular segments of the fruit fly—the head segment that produces a pair of antennae and one of the thoracic segments (say, the middle one) that produces a pair of legs. These two pairs of appendages have, at least in broad terms, the same cell types. For example, they both have the cells that produce the jointed exoskeleton that sheathes them; they both have muscle cells, connective tissue cells, nerve cells, and so on.

What is far more different between antennae and legs than their arrays of differentiated cell types is the gross spatial patterning of the structure concerned. Legs are, appropriately, more heavy-duty structures—after all, an insect does not support its weight on its antennae, whereas its legs are used for just that purpose. Because of such antenna-leg differences in pattern formation, we have no difficulty in distinguishing the two structures, even if they were removed from the fly so that we did not know which segment they had been attached to.

Although there are no mutant cuboid humans, as we discussed earlier, there are indeed mutant fruit flies in which the segment that normally produces antennae produces legs instead. Such flies have four pairs of legs—the normal three pairs produced by the three thoracic segments plus a fourth pair, projecting anteriorly from the head. This mutant form was, for obvious reasons, given the name antennapedia. It is a strange sight to behold.

In fact, these strange mutant flies are only one of several examples of a particular class of mutants, all of which are characterized by having what is, in a sense, the right thing in the wrong place. Another example is a mutant fly with an extra pair of wings. (Flies, unlike many insects, normally have only one pair.) This arrangement is due to the conversion of the third segment of the thorax, which normally carries only very small flight-balancing appendages called drumsticks, into a repeat of the second thoracic segment, which produces wings. This mutant is called bithorax, even though this is a rather inaccurate label since the pair of wings is duplicated but the thorax is not.

These two mutants and others of their kind are unusual. Usually, when something goes wrong in the developmental process, a structure is malformed or absent rather than turned into something else that is normal but out of place. For example, many mutations in fruit flies result in the development of markedly reduced, and often nonfunctional, wings. In recognition of the thing that antennapedia, bithorax, and other mutations of this kind have in common, and by which they differ from the great mass of other mutations, they are given a special name: homeotic mutations. We will revisit these in Chapter 13.

The significance of any homeotically mutant fly (or other creature) is that the form it takes as a result of having something wrong with a particular gene tells us something important about what the normal version of the gene must be doing in so-called wild-type flies. This gene would appear to be selecting one out of a series of possible developmental trajectories for the segment concerned. So the genes that can go wrong in this way have been called selector genes. Their role in pattern formation at a big scale is comparable to the role of the genes that, at a smaller scale, send individual cells on particular developmental journeys to become muscle cells, skin cells, or whatever.

So, in a sense, genes control both cell differentiation and pattern formation. But recall that the term "control" is better avoided in favor of "interaction," because it is a two-way (or indeed multi-way) process. Gurdon's experiments with frog embryos showed that genes can be controlled by their cellular environment. Conversely, the work on homeotic mutations in flies carried out by the American geneticist Ed Lewis and others showed that whole populations of cells can be controlled, in terms of their developmental fate, by genes.

This dance, in which the interacting partners of genes and cells produce wonderful and intricate patterns, is itself, on a much longer timescale, taking part in another dance—that of evolution. In humans, certain genes are clearly counterparts of the fly genes we have just discussed. Their DNA sequences are sufficiently similar that it looks as if they have evolved from common ancestral genes in a common ancestral animal. Moreover, certain aspects of their functions are also similar, despite the grossly different structures of humans and flies. For example, the human equivalents of fruit-fly selector genes also influence pattern formation; not only that, but they also influence segment identity, albeit our own segments (involving vertebrae, ribs, and associated muscles) are less visible from the outside than those of an exoskeleton-clad insect.

The parallels between humans and flies go further still, into the realm of mystery—in the sense of as-yet-unsolved problem, rather than magic. The linear arrangement of the selector genes along the appropriate chromosome in flies is the same as its human equivalent. And in both cases, the anterior-posterior positions of the segments that each such gene influences follow the same order as their genes do on the chromosome—the two things are said to be colinear. This fascinating finding applies also to the vast majority of other animals. Its explanation is not yet clear.

As we are beginning to stray from the developmental realm to its evolutionary counterpart, it is appropriate to end this chapter and begin the next. But in books, as in life, divisions between parts are often less clear-cut than they might at first seem. The dance through which cells and genes create organisms and the longer-term dance in which those partners are deflected in their interaction by natural selection are forever intertwined. Without evolutionary advance from the exclusively unicellular world of a billion or so years ago, there would be no such thing as embryogenesis and other multicellular developmental processes like metamorphosis and regeneration. And without a developmental process to work on, evolution cannot turn one kind of multicelled creature into another, for there is surely no way that a new kind of adult creature can be made other than through an altered developmental journey.

CHAPTER EIGHT

LIFE THROUGH TIME

Although life has graced our planet with its presence for more than three billion years, animals did not arise until sometime between one billion and half a billion years ago. So there was a very long period in which the complexity of creatures changed little. The most important event in this period was the invention of a more complex kind of cell. Once animals appeared, organismic complexity began to rise rapidly. But, as ever, we should remember that hidden behind a rising average complexity, some animals remained relatively simple.

TEACHERS of biology often use a memorable, and, as it turns out, surprisingly accurate, metaphor to describe the comparative brevity of human habitation of Earth. It goes something like this: If the history of life is represented by a twenty-four-hour period starting at 0001 in the wee hours of the morning, then the first humans appeared at 2359. That is, we have only been here for approximately 1/1,500 of the time that life in general has been here. We are, if you like, new kids on the block.

Memorable, yes. But what about my claim that this metaphor is also tolerably accurate? Well, to a rough approximation, life began about 4 billion years ago (BYA). The human lineage diverged from that of the chimpanzee sometime before 5 MYA (note the thousandfold diminution of the timescale to millions), so by 4 MYA there was a human—or protohuman—lineage. As you can see, the calculations are easy in this case, and the human lineage has been separate from that of the apes for about 1/1,000 of evolutionary time.

As ever with statements about life, there are many ifs and buts that could enter into our deliberations. Let's look at just one of them. If you would prefer to consider the origin of humans to be the appearance of *Homo sapiens* rather than some of its ape-man ancestors, then the fraction of life's history for which we have been around reduces to an even smaller fraction—perhaps as little as 1/20,000. While this fraction is very different from the previous ones, it merely serves to rub in their point of our relative recentness on the evolutionary scene.

At that unspecified but early stage in my education when I first encountered the clock metaphor, I for some reason made the assumption that Earth had existed for a comparably short fraction of the duration of the universe. Although I am no cosmologist, I have at least by now picked up enough of that awesome subject through the literature of popular science to realize that this was an extrapolation too far. In fact, Earth is, at 4.5 billion years, about a third as old as the universe itself. So is life, given that its origin was not long after Earth's.

I've digressed a little and will return to a more terrestrial perspective. But I thought it worth drawing attention to the fact that life is not such a latecomer to universal time as we humans are to evolutionary time.

Okay, then, back to planet Earth, and the long march of living things from four billion years ago to the present. We need to look for interesting events and patterns in this long march, and to focus on them, for otherwise the story will be interminable. And interest, in this context, will be determined by relevance to our core mission of trying to understand the evolution of complex creatures.

Another way of making the task manageable is to make comparisons between widely separated points in time. I'm going to mark out my time frame in units of half a billion years. So we will look at the situation of life at the following nine historical markers: 4, 3.5, 3, 2.5, 2, 1.5, 1, 0.5, and 0 BYA. One important, if obvious, caution at the outset: the earlier the time point, the less reliable the information. We have no fossils from 4 BYA, while we have an embarrassment of riches at 0.5 BYA; and at time zero we can see all the magnificent splendor of life for ourselves, rather than having to infer it from fossilized remains in the rocks.

Four billion years ago, there were probably no true cells. This was the approximate time of the so-called primordial soup, in which large molecules formed aggregates, some of which had the accidental property of growing and then splitting—perhaps the first process on Earth that could be considered a sort of life cycle. But a life cycle is all that is needed (along with inheritance of its features) for Darwinian natural selection to come into play. Those aggregates that cycled more reliably would persist over longer periods of time than their more unstable counterparts. So any features that increased the chances of survival and reproduction would tend to prevail. Doubtless one of the most important of these features is some sort of boundary layer that protects the inner workings of the aggregate from environmental buffetings that may destabilize it and stop the lineage in its tracks. If so, then the origin of proto-cells with rudimentary outer membranes is readily understood, and represents the first major step up the ladder of complexity.

By 3.5 BYA, it is clear, even from the very limited fossil evidence we have, that such proto-cells were flourishing. Of course, we have no idea how different their internal workings were from those of present-day bacteria. But in outward appearance they were quite similar.

Strangely, at 3, 2.5, and possibly even 2 BYA, the state of life on Earth, at least in terms of organismic complexity, was not very different. Simple cells were all that existed. Although they sometimes attached together in very basic ways to form one-dimensional filaments, two-dimensional mats, or even three-dimensional stacks of mats, it seems they were incapable of forming themselves into complex multicellular bodies. Perhaps this was because the nature of the cells concerned was such that they constituted too simple a type of building block. Perhaps they lacked one or more of the essentials required to render them suitable for coming together into the metaphorical castle of the multicellular creature. In any event, nothing much happened, in terms of increases in complexity, for at least one, and perhaps even two, billion years—that is, up to half of life's history.

About 2 BYA—estimates vary somewhat—a new and more complex type of cell arose: the kind we and all our animal cousins (and plants and fungi, too) are composed of. This is the eukaryotic cell with a wealth of internal structuring that is absent from the simpler cells of early life and today's bacteria. Most important, each of these cells has, as we saw in Chapter 5, a central nucleus containing the genetic material, together with other membrane-bound organelles with specific functions, such as producing energy and, in the case of plants, photosynthesizing.

It is now generally agreed that this new and more complex type of cell arose from symbiotic relationships between simple cells wherein some would live inside others and a sort of division of labor would evolve. One particular observation on present-day eukaryotic cells strongly supports this theory. Some organelles have their own little genome in addition to the cell's main genome that resides in the nucleus. This looks like a vestige of once-independent existence.

If the origin of simple cells from less organized molecular aggregates represents the first step up the ladder of complexity, then the origin of eukaryotic cells represents the second. Their separation in time by almost two billion years is staggering. Why the long wait? These events are so deep in the past that we may never know the answer. All we can do is state the obvious: if it took that long for eukaryotic cells to appear, then the sort of arrangement on which they are based, involving a particular form of symbiosis of simpler cells, must have been a highly improbable one.

Now follows another long period of stasis as far as the complexity of creatures is concerned. At 1.5 BYA, and probably also at 1 BYA, unicellular eukaryotes were the pinnacle of evolutionary achievement. I suppose that, as with their bacterial predecessors, some formed themselves into simple arrangements like linear filaments, but no impressive "castles" of cells were to be found.

The statement that this was indeed still true one billion years ago would have been accepted without question by the vast majority of biologists as recently as a decade ago, for the simple reason that we had then, and for that matter we have now, no fossil evidence of complex multicellular life from that time. But the once-consensual view on this issue has been replaced, in recent years, with controversy.

The reason for our switch from agreement to argument is to be found in the appearance of a new way of estimating the age of lineage divergences that is quite distinct from, though paradoxically also ultimately dependent on, the fossil record. This new method involves comparing the DNA sequences of two (or more) living creatures, and using the degree of difference found to infer how long ago the last common ancestor of the species being compared lived.

In a bit more detail, it works something like this: You start with a pair of creatures whose parting of the ways is already quite precisely dated from fossils (hence the dependence noted above). Take, say, birds and crocodiles, whose lineages diverged about 200 MYA, the bird lineage having derived from a group of feathered dinosaurs. You look at the percentage difference in some representative stretch of their DNA—perhaps a sequence of a few thousand of those DNA building blocks, the nucleotides.

DNA sequences are thought to diverge in a reasonably regular, clocklike way when considered over a long enough period of time to allow little bursts of rapid change and interludes of relative quiescence to cancel each other out. If this is so, then it should be possible to use information of the sort "X percent difference over each 100 MY" to estimate the date of an older, bigger-scale lineage divergence for which there is no relevant fossil evidence. For example, birds and crocodiles are both members of one of the two great groups of animals we saw earlier—the mouth-seconds in embryological terms. When did this group diverge from its mouth-first counterpart?

In 1990 or thereabouts most biologists would have replied, "About 600 MYA." But when the molecular clock advocates got to work on this question, some of them came up with an answer that was about twice as great—1,200 MYA. It isn't yet clear which camp is correct, or indeed whether the truth lies somewhere in between these highly divergent views. The lack of any believable animal fossils from around a billion years ago counts against the molecular clock estimate. Some clock proponents have tried to explain away the lack of fossils from this time zone in a variety of ways—for example, the creatures concerned may have been very small and/or may have lacked readily fossilizable hard parts like shells, bones, or teeth. But since unicells from 3.5 BYA can provide us with fossil evidence of their existence, this explaining-away argument seems rather weak.

Although the jury is still out on this important issue, I'm going to take the view that at least the very earliest of the molecular estimates of the timing of the origin of animals and of the divergence of their two main groups are wrong. I'll be agnostic about whether these evolutionary events occurred in the 600 (plus or minus 50) MYA region, as was traditionally thought, or whether they happened somewhat earlier (say, 750 MYA). In any event, such agnosticism does not affect the fact that by 500 MYA, the next and penultimate of my nine half-billion-year temporal milestones, animals were abundant and diverse, albeit only in the sea. Colonization of freshwater and land habitats came later.

Geologists and paleontologists divide the history of the earth up into intervals of time called periods. Each of these intervals is tens of millions of years long—exactly how many tens varies a bit between one period and another. Generally, I'm trying to avoid their names, as they're not necessary, and are hard to remember, as there are quite a few of them. But there is one whose name I simply cannot avoid, as it is so crucially important in relation to the history of animal life on our planet. This is the Cambrian period (named after Wales, the Welsh name for which is Cymru, because many rocks of Cambrian age are found there). It lasted approximately 50 million years, from about 540 to 490 MYA. It is the first period in which there are abundant fossil creatures that are indisputably animals.

Incredible as it may seem, almost all of today's major animal groups can be found not just from the Cambrian but from its early and middle parts—before 500 million years ago. A few places in the world have become famous because of their abundant Cambrian fossils. The most famous of all is a location in British Columbia, in the far west of Canada, where there is a particular assemblage of fossils known as the Burgess Shale. These fossils are unusual in that both hard and soft parts have fossilized, thus revealing a level of structural detail rarely seen in extinct creatures. The Burgess fossils, together with material from approximately contemporary sites elsewhere in the world (notably China), include many kinds of animals familiar to us from today's fauna: worms, mollusks, crustaceans, many other kinds of invertebrates, and, perhaps most remarkable of all, some primitive vertebrates, too.

Views on the rapidity of the origins of these groups of animals vary widely, and this variation connects with belief, or lack of it, in the validity of the molecular clock estimates of the age of the animal kingdom. The clock-skeptics think that most animal groups appeared in the blink of a geological eye around the beginning of the Cambrian, and so label this evolutionary event the Cambrian explosion. The clock-advocates think that the explosion of fossils reflects not the rate of evolutionary change but rather the rate at which it becomes visible to us, in which case it is essentially an artifact.

Personally, I am more a skeptic than an advocate of the molecular clock, at least in terms of dating very early evolutionary events. I suspect that DNA sequences may have evolved more rapidly, as multicellularity originated and began to diversify, than they did later on, when their rate of evolution may indeed have settled into a quasi-clocklike regularity. If so, then extrapolation from relatively recent evolutionary splits (like bird versus crocodile) to more distant ones (for example, protostome versus deuterostome) will produce erroneous estimates of the age of those early events.

Of course, I and other clock-skeptics may be wrong. But at least everyone agrees on one thing: by 500 million years ago, almost all the major groups of animals had arrived. So the period of time leading up to the Cambrian can be considered the third step in the complexity of creatures, though I suppose we should really think of it as a flight of steps that some people see as incredibly steep, others as very gentle.

Has there been a fourth major episode in the increasing complexity of creatures? I think not. Rather, there have been small (by comparison with what went before) increases in complexity in some lineages, stasis or even decreases in complexity in others. So if we take a halfway point between 500 MYA and the present, and ask if the overall fauna existing then was more complex than its earlier Burgess Shale counterpart, the answer would seem to be, "Yes, but not much"—albeit some particular increases in complexity in some lineages, for example, the origin of legs, were very important for the creatures concerned. And if we compare the animals of 250 MYA with those of today, the picture would be much the same: specific increases in complexity in some lineages (one of them, increased brain complexity in humans, having massive consequences, and being a focus of attention in Chapter 17), but not much change in the overall picture for the animal kingdom.

It's time for a few caveats. First, we need to be clear about exactly what we are comparing between any two well-separated points in time. It is certainly not minimum complexity, because that has remained approximately constant through most of evolution. When complex eukaryotic cells arose, many lineages remained of the simpler (prokaryotic) cell type, as represented by today's bacteria. They not only remained; they prospered. The same kind of pattern occurred in the later evolutionary event in which multicellular animals arose, because the single-celled eukaryotic creatures that were our ancestors remained and prospered, albeit not to the same extent as bacteria. So our comparisons have really been of three things that almost certainly have shifted in parallel through the main upward steps mentioned above: maximum complexity, average complexity, and range of complexity.

Second, you may have noticed that in telling the story so far, I have flip-flopped between "creatures" and "animals." To some extent, this is simply logical and necessary: you can't talk about animals before there were any. But equally, to some extent it is bias rather than logic, because, in relation to the later stages of evolution, I have been studiously avoiding mention of that other great multicellular kingdom, the plants (and also the fungi, a good many of whose members are multicellular).

Has the complexity of plants evolved in a parallel way to that of animals? In broad terms, I think the answer is yes. But there's also an important difference between the two. The high-level group of animals that includes those with the greatest structural complexity (number of cell types and organs) is the vertebrates. As we have seen, these were already in existence in the Cambrian, by 500 MYA. But the same cannot be said for the plant equivalent of the vertebrates—the flowering plants (or angiosperms, for those in the trade). These advanced plants came on the scene much later—perhaps around 200 MYA. Whether this contrast tells us something important about the difference between animal and plant evolution, or whether it is just a one-off historical accident, is not clear.

Third, there is an enigmatic group of creatures from the period immediately before the Cambrian that I have not yet mentioned. They are sometimes called the Ediacaran fauna—though Ediacaran biota is preferable, since their exact nature remains obscure. The name comes from the Ediacara Hills in Australia, which is where fossilized remains of these creatures were first discovered, though they have since been found in many localities spread across several continents.

These creatures have been interpreted in at least four different ways: They are early representatives of some of today's main animal groups, in which case the Cambrian explosion, if it was explosive at all, happened earlier than the base of the Cambrian period, and so is misnamed. Alternatively, they were a failed experiment in animal design—faunal rather than floral, but not belonging to any of our current animal groups. Another possibility is that they were a strange group of marine lichens. Finally, they may have been an altogether separate multicellular kingdom that arose from the realm of the simple quite independently from animals, plants, and fungi. It is even possible that some of the constituent creatures of the Ediacaran biota are correctly interpreted in one of these ways, some in another.

I'm approaching the end of my "Life Through Time" story. There are just two remaining connections to make before moving on to the next chapter.

Most tours through the history of animals, or the history of life more generally, are centered not on changes in complexity but rather on patterns of relatedness—that is, they focus on elucidating who is related to whom, and to what degree. My alternative focus should not be interpreted as implying a disregard for the issue of relatedness. This issue is of much importance: changes in the complexity of creatures (and in their structure, function, life cycle, and behavior more generally) must be seen against the backcloth of "the true tree of life," inasmuch as we are able to discern it. My reason for omitting a major discussion of this issue is simply that so many such discussions are already available elsewhere.

Finally, in dealing with increases in complexity over evolutionary time, I have been implicitly concentrating on adults. If we concentrate on earlier stages of development, the increases in complexity are reduced, and may even disappear. After all, given that most animals begin life as a fertilized egg, comparisons of the beginnings of life cycles rather than their ends reveal no significant changes over the last two billion years, and permit a view of the present-day fauna as a world of unicells. To be sure, this is a rather strained (and strange) view, but it does serve to remind us that what has evolved is the whole developmental trajectory, not just its end product, the adult. And that trajectory, in complex creatures, re-creates evolution's laboriously slow increase in complexity, at lightning speed, in every generation. The two great creative processes of life, evolution and development, are intrinsically linked, despite their vast differences of timescale, as we saw earlier. We will now inspect the way in which this linkage becomes manifest when the embryonic rather than adult forms of different animals are compared.

CHAPTER NINE

THE EMBRYO WARS

Fish embryos have a series of parallel lines etched into the sides of their heads. These gill clefts will, as their name suggests, later develop into gills. We humans, when we are embryos, also have gill clefts, but clearly these do not go on to develop into gills, as we are air-breathers and use lungs instead. So why do our embryos have gill clefts? This question has long fascinated biologists. Theories in this area, especially the one called recapitulation (development recapitulates evolution), have been deeply controversial. Here we examine this controversy, and note that in the end the "embryo wars" have given way to peace.

THIS part of our story is firmly rooted in Germany. It starts with attempts by the nature philosophers of old to elucidate the pattern of relationship between the series of embryonic stages found in any one kind of animal and the array of different kinds of adult animals found in nature. It then continues through criticism of nature philosophy to an evolutionary interpretation of it. There were many key players, but I am going to restrict my streamlined version to just three: Johann Meckel, Karl von Baer, and Ernst Haeckel. These players wrote their most important works in historical sequence, albeit their lives overlapped. The three key years were 1811 (Meckel), 1828 (von Baer), and 1866 (Haeckel). Notice that the first two of these fall before, and the third after, that supreme historical landmark in the literature of biology, the publication of Darwin's *On the Origin of Species* in 1859.

Now, you might be forgiven for wondering, at this point, just how relevant such ancient works—written one and a half to two centuries ago—are in today's Age of the Gene. But wonder no more. They are supremely relevant. Genes are only part of our overall picture of evolution. They have added to, not replaced, earlier-established types of information, notably, in the present context, information on the physical appearance of the embryos of different kinds of animals. Evolutionary change from one animal to another involves changes in the course of development and in the genes that control it. These are different sides of the same coin. There has been a rekindling of interest in comparative embryology in recent years, and this is very much built on the foundations laid by von Baer and Haeckel, just as recent work on natural selection is built on the foundations laid by Darwin.

But our present story begins in pre-evolutionary times. Starting sometime around 1790, various German thinkers, and a few counterparts from elsewhere, notably France, began to develop a big idea: that all animals were cast in the same mold, and that some were just more elaborated in their development than others. There was a natural scale, as we saw in Chapter 1, and any animal could be placed somewhere on this scale—above some of its fellow creatures, below others. Well, almost any animal. Humans were unique, in this view, because they were at the top of the scale, with all other creatures beneath them.

Meckel was one of the leading nature philosophers, and he, like several others, stated that each animal, in the course of its development, passes through stages resembling the adults of lower animals. For example, take the development of the human heart. At a very early stage of development, it is just a sort of tube. This kind of heart is found in a plethora of lower animals, including insects. At a later stage, the human heart is more substantial, and consists of two chambers, like the hearts of fish. Later still, it has three chambers, like the hearts of most adult reptiles, before acquiring its final, four-chambered form that is familiar to high-school students of biology (two atria and two ventricles).

Now along comes a problem. At no stage does a human embryo look, overall, like an adult fly, salmon, or lizard. Even when attention is focused on particular organs or parts, the resemblance with the adults of lower animals is often very imperfect. It has long been known that early human embryos possess a series of lateral clefts in their head regions. These look, for all the world, as if they will develop into gills, but they do not. So these structures in early human embryos resemble not the gills of adult fish but rather the structures in an embryonic fish that will develop into gills at a later stage in the fish's developmental journey.

Subsequent players in the story refined the picture in just this way—to refer to the resemblance between the embryos of higher animals and the embryos of lower ones. But before we turn to this more reasonable version of the pattern, let's briefly consider the issue of mechanism. So we ask: What is the underlying cause of the pattern?

One of the main problems with nature philosophy was that its proponents were really not too concerned with mechanisms. They were romantic thinkers, idealists of a certain kind, in whose worldview universal patterns were central, whereas the means by which those patterns were generated were not. These folk were content to think in terms of a vital force, of which some creatures had more than others. Those with more of it got propelled further along the route of the universal developmental pathway. Those poor unfortunates with less simply stopped at an earlier point.

It's a bit like taking one particular London tube line, say, the Victoria Line, extracting it from the overall tube network, and pulling it at the ends so that it lies straight in the vertical dimension. All animals start their developmental journey at the most southerly station (Brixton), but only humans get to the northerly top of the line (Walthamstow). Insects stop at Vauxhall, fish at King's Cross. Reptiles get to Finsbury Park, and chimpanzees, so near yet so far, get to penultimate Blackhorse Road. Perhaps the different amounts of vital force can be likened to tube trains powered not by electricity but by diesel fuel. The insect train starts with its tank nearly empty; the fish train starts half-full; and so on.

This picture, of course, is nonsense, for at least two reasons. First, there is no single developmental pathway; there are many. Second, there is no vital force: the vitalist interpretation of life has been firmly defeated by its mechanist counterpart based on physical and chemical laws rather than mysticism.

Enter Karl von Baer. A careful and meticulous student of comparative embryology, von Baer knew very well that the facts of development were not in keeping with the romantic views of the nature philosophers—a camp in which he was schooled but from which he later defected. Development had many possible routes, not just one. And, related to this, the embryonic stages of any higher animal were never exact copies of the adult forms of lower ones. Indeed, they were often nowhere near.

Von Baer developed a series of laws that embodied his overall vision of comparative embryology and that can be seen as an anti-nature-philosophy manifesto. His mental picture was not of a single line of development but rather of a series of lines that radiated out from a common starting point. Some lines ran in parallel for a short distance, others for a longer one, before they parted company. This overall view can be referred to as von Baerian divergence. His laws describe various aspects of this picture. In particular, one of them explicitly denies that the embryos of higher animals resemble the adult forms of lower ones.

So far so good. But while von Baer solved one of the problems inherent in the nature philosophers' view (the line), he did not solve the other (the lack of mechanism). Bizarre as it now seems, von Baer, a man with what many of us would regard as the most persuasive evidence of evolution at his fingertips, did not believe in evolution, either when he published his magnum opus in 1828 or, in later life, after Darwin's publication of his theory of evolution by natural selection. Indeed, toward the end of his life, von Baer wrote several essays attacking Darwin's theory.

It was left to Ernst Haeckel to champion a Darwinian comparative embryology. Haeckel coined the so-called biogenetic law, the most succinct form of which is that "development repeats (or recapitulates) evolution." That is, the developmental journey that a particular type of animal takes within its lifetime is a condensed version of the much more protracted evolutionary journey that its ancestors collectively took to arrive at that particular type of animal.

Now here comes the crunch: Is this "law" compatible with the nature philosophers' view or von Baer's? If the former, as some have argued, Haeckel had merely taken one step forward and one back.

But this is an erroneous interpretation of Haeckel's work. The key question here is whether Haeckel saw the embryos of higher animals as passing through stages resembling ancestral adults or ancestral embryos. The first of these is mystical nonsense; the second, as we shall see, is a predictable consequence of Darwinian evolution.

A few years ago, I sat through a talk at a major international conference where the speaker waxed lyrical about his conviction that Haeckel saw adult ancestors in the embryos of higher animals. In the course of this talk, my complexion probably turned from pale to pink to purple. At question time, at the end of the talk, I asked whether, during the "molluskan" stage of human development, we were garden snails complete with shells or octopuses complete with eight elaborate tentacles. Suffice it to say that I did not get a satisfactory answer.

The linguistic education that most of us in the English-speaking world receive is woefully inadequate. We are, quite simply, spoiled. The rest of the world speaks our language, so why should we bother to learn theirs? Yes, I did my compulsory years of elementary French, but no, I cannot hold a reasonable conversation in that language with a French colleague. I even did a year of German, the end result of which is that now, some forty years on, I can count to ten in German, order a beer in German, and understand, without the services of a translator, what John F. Kennedy was getting at when he said *"Ich bin ein Berliner."* But as for reading German books, I am ill equipped indeed.

The connection between this digression and the point about the nature of Haeckel's worldview is that his magnum opus, like von Baer's, was published in German. There is, to my knowledge, no complete English translation of it available. Perhaps the task was too daunting to anyone who contemplated it: Haeckel's book runs to two volumes, together comprising more than a thousand pages. To truly know what Haeckel thought, and whether and how his ideas changed over his lifetime, we would need to read this whole book *and* all his other publications, too.

Although I cannot do this, and so cannot rule out the possibility that at some point in his lengthy career Haeckel said something that could be interpreted as meaning that human embryos have an adult-octopus stage, I do at least have one insight into Haeckel's writings that leads me to believe he thought nothing of the kind. This is because a later book of his, which first appeared in 1874 (as *Anthropogenie*), was translated into English in 1896 as *The Evolution of Man*.

There is a magnificent passage early in this book that makes Haeckel's view on the adult-versus-embryo issue abundantly clear: "The fact is that an examination of the human embryo in the third or fourth week of its evolution [development!] shows it to be altogether different from the fully developed Man, and that it exactly corresponds to *the undeveloped embryo-form* presented by the Ape, the Dog, the Rabbit, and other Mammals, at the same stage of their Ontogeny [development!]."

One sentence out of thousands—yet what a sentence! It is heavy with meaning, yet also potentially misleading. A little dissection is appropriate. First, ignore the synonyms for development, especially "evolution," which, confusingly, was indeed used in that way. Second, ignore the idea of an exact correspondence—Haeckel probably knew, and we certainly know, that the correspondence between, say, a three-week human embryo and a three-week dog embryo is impressive but inexact. Now, finally, focus on the bit that I have italicized. Haeckel saw no *adult* ape, dog, or rabbit in the early human embryo—of course he didn't. So he did not take a step back into the nature philosophers' camp; rather, he was a staunch von Baerian.

Why, given this fact, is Haeckel so often misinterpreted? Perhaps the biogenetic law, and in particular the idea of recapitulation, seemed to hark back to the nature philosophers' way of looking at things. But this is mere illusion. Haeckel's ability to think in terms of both von Baerian divergence *and* an enlightened form of recapitulation was due to his having something that von Baer did not: a belief in evolution. Von Baer only thought in terms of comparing one living animal with another. He did not think in terms of ancestors, because he did not believe in them. Haeckel, on the other hand, was able to think about ancestors as well as their present-day descendants. Divergence of the embryonic trajectories of two living forms is quite compatible with recapitulation, especially in the case where an evolutionary split has been accompanied by a complexification of one of the lineages. The embryonic trajectory of the more complex form, in such cases, will recapitulate, albeit imperfectly, the trajectory of its simpler ancestor, before moving on to greater things.

And so to Charles Darwin. First and foremost, Darwin was a naturalist. His own personal observations that drew him to his theory of natural selection were made not on embryos but rather on later developmental stages, including adults. These he saw in their natural ecological context in the wild all over the world, as HMS *Beagle* stopped at diverse places, including, most famously, the Galápagos archipelago off the coast of Ecuador.

But while Darwin was not an embryologist, he was by no means ignorant of that important biological domain and its relevance to his theory. He devotes little space to it in *On the Origin of Species* (a small part of his Chapter 13), but what he says there is, as you might expect, spot-on. He points out that the variations on which his theory depends manifest themselves at various developmental stages, some earlier, some later. Now, if natural selection favors a variant that appears at point X in embryogenesis, and employs that variant in the course of a divergence between lineages, so that one lineage is characterized by it, the other not, then a comparison of pre-X embryos of the descendant species will show similarity, while a comparison of post-X embryos will reveal a difference.

A single difference is hardly realistic in a comparison of two species whose lineages diverged many millions of years ago—let's say humans and fish. In such cases, many variants, manifesting themselves at a variety of developmental stages, will have been naturally selected. Since changes in early embryos will often cause subsequent changes in later ones while the converse is not true, evolution by natural selection would be expected to produce von Baerian divergence and, in cases of rising complexity in one lineage, Haeckelian embryonic recapitulation.

Darwin was a modest man and a beautiful writer. Toward the end of his short section on embryology, after rehearsing the various arguments about the relationship between evolution and development, he sums up in a single, short, understated sentence: "Thus, community in embryonic structure reveals community of descent."

For those who like neat stories, the picture we have arrived at is rosy indeed. The early nature philosophers, such as Meckel, were wrong, both in their proposed pattern and in their mechanism. The later evolutionists, notably Darwin and Haeckel, were right, again in both. The heroes and the villains are easily separable, except perhaps for von Baer, who got the pattern right but not the mechanism, as he refused to believe in evolution.

As the Oracle says to Neo in *The Matrix*, I hate giving good people bad news. However, the neat story is not so neat after all. History did not end in 1866, or even 1966; indeed, perhaps history has no end. In any event, our story continues and the water muddies . . .

Since the 1980s, a three-pronged attack has been mounted on the consensus that I have just described. The first prong involved recognition of the fact that embryos of different animals have maximum similarity not right at the start of development—the fertilized egg—but rather a little later. Thus, if we compare the developmental trajectories of several animals, all of about equal complexity, we see not a simple pattern of von Baerian divergence but rather convergence followed by divergence.

This point was first made by another German, Klaus Sander, in the context of insect embryos. But it was immediately clear that the point was valid for a much wider range of animals than insects. For example, the early embryos of birds and mammals are very similar, but their eggs could hardly be more different. The very earliest embryonic stages are thus constrained to be different because of their different situations. The convergence-followed-by-divergence pattern has been referred to, for obvious reasons, as an hourglass or egg timer. In terms of accuracy, I'd have to admit that this is preferable to its von Baerian predecessor. And yet I find my worldview little altered by it. I think this is because, while a real egg timer is symmetrical, its comparative embryological counterpart is about as asymmetrical as they come. If, in a comparison of a human embryo with that of another species of vertebrate, I have to acknowledge that the point of maximum similarity is at week 1 rather than week 0, that doesn't seem like a big deal given that our embryogenesis takes about forty weeks (and our overall development nearly twenty years).

The second prong was the gradually growing recognition that the comparative embryology of the nineteenth century had been too dominated by studies of mammals (especially), birds, and fish, all of which develop directly via juveniles rather than indirectly via larvae. But consider indirect developers such as frogs, butterflies, and sea urchins. In each case, the larva and the adult are so different that a major metamorphosis is required to get from one to the other. Moreover, the larva and the adult differ ecologically as well as morphologically. Frogs move from water to land as they develop, butterflies from land to air, and sea urchins (whose larvae are so small that they are unknown to the general public) from plankton to seabed.

Von Baerian divergence, whether in original or modified (egg timer) form, is clearly an inappropriate way of looking at the whole of development in a comparison of two or more such creatures. However, again I find my worldview relatively unaffected. Why?

Imagine the developmental sequences of two butterfly species from different families lined up side by side. In many cases, similar early embryos give way to very different caterpillars. Then, hidden from view inside their opaque chrysalises, similar metamorphosing pupae, with similar rudimentary wing buds, give way to very different adults, each with its own striking wing pigmentation pattern. So, instead of having one divergence, or egg timer, we have two, joined end to end.

The third prong of the attack came from the English biologist Michael Richardson, who suggested that the drawings of early embryos made by Haeckel had been deliberately falsified so as to reinforce his point. Actual differences between the early embryos of various types of vertebrates had been downplayed to the point where they were less obvious or even invisible. Again, I find my worldview unaffected.

Of course, scientific fraud is a terrible thing. It is a matter of honor among scientists of all kinds that we do not distort our findings in order to bolster our theories. And yet fraud is not uncommon. It appears that Gregor Mendel, the nineteenth-century monk who founded the science of genetics, may have doctored the numbers of different kinds of progeny in the results of his experiments with garden peas in order to give extra credence to his theories. If true, this reduces his integrity in the eyes of subsequent scientists, but it does not stop us from believing in his fundamental laws of inheritance. It seems a regrettable thing to those of a purist disposition, but fraud needs to be considered in context. If fraud makes a pattern appear less fuzzy, but the pattern is nevertheless real, it is a pity. If, on the other hand, the fraud urges acceptance of an illusory pattern as if it were a real one, then it is a crime.

I have got almost to the end of this chapter with barely a mention of complexity. It has been implicit rather than explicit. That's true of much of the book, of course—the complexity of creatures is always with us as a sort of invisible escort, even when we are straying a bit from our main theme. But now is a good time to bring it into full view again, in order to be crystal clear about the fundamental message of this chapter.

So, back to Darwin; and specifically, back to that sole source of embryological wisdom (Chapter 13) in his *On the Origin of Species* to which I referred a short time ago. Darwin acknowledged that development itself is normally a process of increasing complexity—though instead of "complexity" he used "organisation." He says: "The embryo in the course of development generally rises in organisation: I use this expression, though I am aware that it is hardly possible to define clearly what is meant by the organisation being higher or lower." And, true to form, he notes that although this is the rule, there are also exceptions—especially in those parasitic forms where the adult is simpler than some of the stages that lead to it.

Fast-forward to 1977 and we find another old friend, Stephen Jay Gould, making the same point, despite his mischievous downplaying of complexity with which I started this book: "'The development of complexity during ontogeny [development] . . . and the recognition that there are 'higher' and 'lower' species are two inescapable phenomena of biology."

So development (normally) leads to greater complexity, and evolution (on average) leads to greater complexity, and these two ascents of life's ladder are related. Creatures whose ancestors have struggled upward through the generations must themselves struggle upward through the formative phase of their life cycle. Their journey is not the same as that made by those distant ancestors, but it has many features in common. Such a compressed and imperfect "recapitulation," to use that word that has caused so much trouble, is inevitable, because of the way Darwinian evolution works. Developmental trajectories get modified piecemeal, not replaced wholesale. You cannot expect a "high" animal to burst preformed in all its complicated glory out of a minuscule egg. Such is the stuff of cartoons or fairy tales. This is a case where the course of human thought has been much hampered by the carrying of heavy philosophical, indeed even political, baggage. Extreme versions of recapitulationism have sometimes been associated with extreme right-wing views. Gould accuses one of the nature philosophers, Lorenz Oken, a contemporary of Meckel's, of being "among the intellectual antecedents of German fascism."

I have no truck with fascism, a form of politics associated both with lack of freedom of expression (even book burning) and with some of the worst atrocities ever committed. But we should not allow our understanding of scientific processes to be influenced by the politics of the advocates of particular views about the nature of these processes, however unsavory those politics may be. We cannot extract science from the social milieu in which it operates, but we should at least attempt to see where the join between them is, and to be wary of carrying messages across it from one domain to the other.

CHAPTER TEN

DARWIN AND HIS LEGACY

Charles Darwin's main contribution to our understanding of the nature of life lay in providing us with a simple and effective mechanism through which evolution could work, namely natural selection. He said a lot about how this mechanism would produce variety, but relatively little about how it could lead to increases in complexity. Here we will focus on the latter issue.

HERE is a supreme lesson in the communication of science. In 1858, Charles Darwin published his revolutionary theory of the evolution of life through the process of natural selection. It appeared as a jointly authored article with Alfred Russel Wallace in the *Proceedings of the Linnean Society of London*. Wallace was the man whose independent arrival at the same theory had forced Darwin into print after many years of quietly sitting on his theory, accumulating further evidence for it before daring to bring it out into the daylight of public view.

You might think that the publication of this joint article would have caused a stir—or even a storm. But no. Not only did the general public not react, but even the Linnean Society's own account of the discoveries published in its journal that year was something along the lines of "nothing of much consequence."

So there we were—humans, that is: the only animal species out of millions to have begun to think and write about its own origins.

And we finally had a possible explanation of how we got here. And to all intents and purposes, it went unnoticed. There were no shrieks of "Eureka," nor cries of "Shame" from objectors. Just silence.

How can we explain this odd result? Not through the unimportance of London as a city—it was one of the world's most prominent then, as now; indeed perhaps even more so, since the British Empire was still a major force in the world and the United States a mere upstart among nations. Not through the obscurity of the Linnean Society, either. Named after the founding father of the classification of life-forms, the Swede Carolus Linnaeus, it was then, as now, a leading scientific society (albeit not London's foremost, a position occupied by the quaintly named Royal Society—really a National Academy of Science in disguise). Not through the insignificance of the language in which it was published (English, of course); and not through the lack of credentials of its authors.

By a process of elimination, we have arrived at what I believe to be the real culprit: the title. This was as follows: "On the Tendency of Species to Form Varieties; and on the Perpetuation of Varieties and Species by Natural Means of Selection." Too long, a bit repetitive, and, most important of all, the wrong way round.

Why the wrong way round? Well, think of it like this: Any naturalist or even gardener or dog breeder of the mid-nineteenth century knew perfectly well that species formed varieties. That was no news at all. But invert it to "On the Tendency of Varieties to Form Species" and you move at once from trivia to heresy.

It would take a better historian of science than I am to unravel the reasons for such an obvious (at least in retrospect) gaffe. As ever, history is complicated. In this case, the complications included the absence of both Darwin and Wallace when their joint paper was read to the Linnean Society (earlier in the same year that it appeared in print). Darwin was at home, where his son was dying; Wallace was abroad. Also, the paper presented under the inverted title was itself a composite of several bits written independently by the two great men (one of which, by Wallace, had a better title than the overall one).

It's a fascinating exercise to try to come up with a title that might have led to a more pronounced response. How about the following? "The Origin of All Species, Including the Human One." "The Mechanism Underlying the Production of All Species on Earth from a Single Ancestor." Or simply: "How All Forms of Life Have Arisen."

It is quite impossible, one and a half centuries later, to know what the reception of the paper would have been if it had been titled in one of these ways. But in the end it matters not, because we do know what happened a year later, when Darwin published his magnum opus, *On the Origin of Species*, which sold out almost immediately. It ended up being reprinted many times and running through several editions—a copy of the first of which, in good condition, is now worth about fifty thousand dollars.

What happened in 1859 was the delayed storm of debate and argument that should have followed the publication of the joint paper in the preceding year. Many of the characters who took part in the debate are famous, especially Soapy Sam Wilberforce, the incensed creationist bishop; and "Darwin's bulldog," the articulate Thomas Henry Huxley, grandfather of another well-known biologist, Julian Huxley, and also of the famous novelist Aldous Huxley.

Nothing in the affairs of society takes the form of an ideal experiment—if it did, history would be easier to interpret than it is. In an ideal experiment, you can draw a firm conclusion about the effect of something (call it X), because the presence or absence of X is the only difference between two otherwise identical experimental treatments. For example, in a biology laboratory, the effect of temperature on the rate of growth of maggots can be tested by growing some in an incubator at one temperature, and others, of the same species and with an identical food supply, in another incubator at an appreciably higher temperature. We know that maggots grow faster at higher temperatures because of the outcomes of just such experiments carried out by competent scientists. If those experiments had been done instead by incompetent scientists, whose two incubators had differed both in temperature and in something else (like the type of food available), then we would have learned nothing.

The contrast between the storm that followed the publication of Darwin's magnum opus in 1859 and the earlier calm that had followed publication of the joint Darwin/Wallace paper in 1858 was, like most historical contrasts, akin to the experiment done by the incompetent scientist. In fact, it was worse, because at least *three* things (and probably many more) were different. First, the title had become reinverted to *On the Origin of Species* (or, for those who like to see titles in full, *On the Origin of Species by Means of Natural Selection; or, The Preservation of Favoured Races in the Struggle for Life*). Second, books are received very differently from papers. Third, a year had passed, in which countless conversations about evolution (and everything else) had taken place, in a complex social dynamic that we can never perfectly re-create or understand.

Which of these three things (or indeed others) mattered the most? You will find many books and learned papers in the literature of what has become known as the Darwin industry that will attempt to tell you the answer. But from the perspective of my own main theme of climbing life's ladders, this has all been a digression. Perhaps, you might agree, an understandable one, because anyone interested in science communication cannot fail to be fascinated by historical lessons in how to succeed (or not) in getting across big ideas to a wide readership. But also, in the end, a limited one, because we need to get back to our main theme.

So how did Darwin's stance connect with the evolution of complexity? Strangely, the answer is, "Very little." That's not to say that Darwinian natural selection is not important in upward evolution—it is. But Darwin did not develop this point. Instead of focusing on complexity, he focused on diversification. That is, life's lawn rather than life's ladders.

We have already seen that many evolutionary case studies, such as that of Darwin's finches, tell us little about the evolution of complexity, because all the species concerned are about equal in this respect. Finches with big beaks are no more or less complex than those with smaller ones; they're just different.

I want to spend the rest of this chapter developing this point that Darwin did not develop: the role of natural selection in evolving upward in particular.

Like so much about evolution, the production of complexity by natural selection is an accident, or, if you prefer, a by-product. Selection is never trying to achieve anything—it just happens. Because of the nature of the selective process, it inevitably produces adaptation. That is, it causes organisms to have forms and functions that are appropriate to their environment. It does not optimize such forms or functions; rather, it makes the best of a bad job in that it increases the frequency of fitter variants and decreases the frequency of less fit ones. Indeed, this can be thought of as a definition of natural selection (or, if you are in a more negative frame of mind, a tautology). Selection can only work with whatever variation it finds. In whales and dolphins, it has managed to turn legs back into fins, but not lungs back into gills.

Most selection, most of the time, is probably of this general kind: making limbs more appropriate to the nature of the environment; making beaks more appropriate to the nature of the food supply; and so on. We might call these changes complexity-neutral. But in some small proportion of cases, the thing that is fitter is also *more* or *less* complex than the thing it replaces. These, especially the "mores," are the cases on which we must focus our attention.

In fact, I'm going to focus on the role of natural selection in climbing life's ladders at two very different points in evolutionary time: "early," let's say about 2 billion years ago; and "late," for example, a mere 20 million years ago (or 0.02 billion, to make the figures more readily comparable). As we will see, the way natural selection connects with rising complexity is different in these two cases.

So picture a superficially lifeless Earth: no trees, no grass, no insects, no people. You are an alien observer who has landed on this "third stone from the sun" (as Jimi Hendrix memorably labeled it) about two billion years ago. You sample the water of the primitive seas, you scrape some slime off a wet rock under a waterfall, and you return to the lab in your orbiting spacecraft to analyze these samples for signs of life. The answer that spews out of your lab's computer is clear: carbon-based life-forms with a typical chemistry for such things (you have found them in other galaxies, too), with enzymic players making one kind of molecule into another in an endless dance of interactions, the various players being fabricated by genes.

Although you have been traveling for many years and made similar findings on many planets, you have not yet lost your sense of wonder and curiosity about what such simple life-forms look like. So you put a slide of slime under the microscope and have a look. You see lots of simple cells, which you recognize as bacteria. But in one remote corner of your field of view, you notice a bigger cell that, on closer scrutiny at a higher power of magnification, can be seen to contain lots of little substructures. Some of these look a bit like bacterial cells being held prisoner within the cytoplasm of the bigger cell.

Your privileged position as the first intelligent visitor to the early Earth has given you a direct view of something that we of the much later terrestrial intelligentsia can only infer from indirect evidence: the origin of complex cells from simple ones. Let's now think about the role of natural selection in this process.

There's nothing wrong with the design of bacteria. Even in today's world, they are the most numerous life-form. For many ecological purposes, simple is fine. There was never any need for natural selection to make creatures more complex. But of course natural selection does not think, it does not perceive needs, it does not plan or anticipate; it merely acts. In the origin of complex cells from simple ones, it may have acted because, in a particular bay off the coast of a particular island, a large bacterium engulfed some smaller ones but, instead of consuming them, retained them as metabolic slaves: internal powerhouses into which could be partitioned some cellular functions, while others continued to occur in the cytoplasm of the host cell.

If the incipient division of labor that this new arrangement permitted resulted in increased efficiency, reflected in increased survival and/or reproduction, the complex big cells would have prospered, and ultimately prevailed over the simple big cells. This is natural selection in action, in just the same way that we saw it act to change populations of pale moths to populations of black ones in urban areas of England, after the soot of the industrial revolution had blackened the trees.

The origin of complex cells some two billion years ago was, as we noted earlier, the first noteworthy step up the ladder of complexity, after the origin of life itself. But recall the nature of this ladder, or, to be more precise, the nature of the thing that is climbing it. Minimum complexity is not climbing the ladder; it is perfectly happy on that bacterial bottom rung. The two things doing the climbing are maximum complexity and (the slower of the two climbers) average complexity.

The rise in maximum complexity is the simpler of the two to think about, so let's deal with it first. If we assume that all the bacteria of two billion years ago were about equally simple, then the origin of the first complex cells takes maximum complexity up a rung. A small step for a cell, a big step for cell-kind, as we might say, borrowing from Neil Armstrong.

But what happens to average complexity is another story. Clearly, it rises—but by how much? That depends. Averages seem like the simplest of all statistical devices. Schoolchildren learn about them at an early stage, long before those that pursue their mathematical education further come across such exotica as standard deviations and confidence limits. But averages are not simple at all. There are often several different ways of calculating them, each of which yields a different numerical result.

In relation to two-billion-year-old cellular evolution, we might want to say that the average complexity of all living creatures is just that: you measure the complexity of every cell on the planet, add the measurements together, and divide by the number of cells. (At this early stage in the history of the biosphere, "cell" and "creature" are synonymous.) To make the calculations easy, let's say that we arbitrarily give a bacterial cell a complexity index of 1, while we give the first eukaryote cell, as the more complex ones are called (literally "true-bodied"), a complexity index of 10. If there are 99 bacteria in the world (a ludicrously low number) and 1 eukaryote, the average complexity has risen from 1.0 to about 1.1: a 10 percent increase.

As the ancestral eukaryote goes forth and multiplies, the average complexity of creatures goes up, so long as this multiplying outpaces that of the simpler bacteria. But if, instead of calculating the average from numbers of individual creatures, we decide to calculate it on the basis of numbers of *species*, then reproduction alone will not affect it. In this case, the average will only increase if further evolutionary changes, specifically the origins of new species, are biased in favor of the eukaryote.

Regardless of which exact method of calculation we use, there is another factor that we have yet to consider, and that will have an effect not only on the magnitude of change of the average but even, potentially, on its direction. In fact, the average complexity could, paradoxically, decrease, at the very moment when the first eukaryotes arose. How?

It all depends on what other evolutionary events are happening in places other than the particular stretch of island coastline at which this discussion began. Suppose that a few hundred kilometers along the coast, another bacterium becomes simpler. This could counterbalance the increasing maximum complexity with a decreasing minimum, thus potentially leaving the average unchanged. If two separate lineages decrease, they will outweigh the single eukaryote increase, and the net change in complexity will be downward.

I'm going to argue that such decreases never happened. There must be a minimum complexity for independent existence as a lifeform. I'm going to make an assertion, and it's only that, that this minimum is represented by a typical bacterium. As a ballpark estimate, I'm actually quite confident about the truth of my assertion, though there are a few ifs and buts, as follows. First, bacteria are not all equally simple, so perhaps we should refer to a minimal rather than a typical one. Second, if a bacterial design were really the simplest possible design, life could never have arisen, because those molecular aggregates and proto-bacteria that started the whole thing off would be prohibited. The solution to this puzzle is that the minimum viable complexity is context-dependent. You can be a loose proto-cell with only a rudimentary outer boundary and no proper cell membrane in a world where you are competing only with other similar life-forms, but in the presence of bacterial competitors you are doomed. Finally, the importance of specifying a capability for independent existence is made apparent by considering those quasi-life-forms that we call viruses. These are much simpler than bacteria, but none of them can exist on their own: all are "pirates of the cell," to borrow the lovely book title by the British biologist Andrew Scott.

If my assertion that, subject to those ifs and buts, nothing simpler than bacterial design was possible two billion years ago, then no decreases in complexity would have been permitted by natural selection. In contrast, occasional increases would indeed have been permitted. So both average and maximum complexity increased with the advent of the eukaryotic cell.

However, this is a unique situation. If there is a floor to complexity, then early changes in complexity that occur when all life-forms are floor-bound are compelled to be upward ones. There is, you might say, nowhere to go but up. But once upward changes have occurred, downward ones must be possible, too. So now we fast-forward to later evolution to consider this more tricky, but more usual, situation.

I deliberately chose twenty million years ago as my example of "late" evolution so that it would not be overwhelmingly influenced by any particular one-off event. If I had chosen sixty-five million, I would have entered mass-extinction territory, the special nature of which I want to leave until later (Chapter 18). On the other hand, if

I had come forward to five million, I might have risked focusing on human evolution, which is again something rather special that again I would prefer to postpone (this time to Chapter 17). In contrast, at twenty million years ago, evolution was routine, if that is not a contradiction in terms.

What happens to the average complexity of creatures during such periods of routine evolution? No one can claim to have a cut-and-dried answer to such a difficult question. Let's start by looking at why it is indeed so difficult.

Twenty million years ago, there were creatures at many levels of complexity, from bacteria all the way through to mammals. From one cell to trillions of cells; and from a single cell type to more than a hundred. Each level of complexity was represented by many species. Over a period long enough for meaningful evolutionary changes to take place—say, the ensuing two or three million years—the following things will happen.

Some lineages will split, creating new ones. Others will evolve without splitting. Still others will become extinct. In the first two cases, the evolutionary changes may be upward, downward, or complexity-neutral. The last of these will probably predominate, yet they will be irrelevant to any shift in the average. This shift will be the net effect of the number and magnitude of upward and downward changes, and of the difference in average complexity of lineages that remain extant and those that go extinct.

I think that last paragraph adequately conveys the difficulty of the issue. And because of this difficulty we should be appropriately cautious in making generalizations. So preface the sentences of the following paragraph with "probably" or "I think that" or whatever other appropriate indicator of tentativeness you prefer.

In some periods of routine evolution, the average complexity of creatures goes up; in some it goes down; and in others it stands still.

Which it does for any one such period is unpredictable. However, in the long run, the net change is upward. This is because fitness is more frequently enhanced by adding new components to a creature than by losing old ones. (I'll expand on this last point in Chapters 12 and 13.)

Darwin's legacy is, in most respects, a positive one. He was an intelligent and honest man who worked hard on an important problem, had a major new insight regarding a possible solution, and (eventually) published this insight in a beautiful and highly readable book. But there is one negative aspect to his legacy: an overemphasis on natural selection as *the* cause of evolutionary change rather than as *one* such cause. Actually, it is not to Darwin himself that we should attribute this negative legacy—rather, it is the fault of some of his hyper-Darwinian followers.

The other cause of the evolutionary process that we need to consider is the origin of the variation upon which natural selection acts. Complex cells may sweep through populations of simpler ones under the influence of natural selection. But where did the first complex cell come from? Equally, black moths may sweep through populations of pale ones, but where did the first black moth come from? Classically, such questions take us into the realm of mutation. But care is needed here, because mutation is just a change in a gene. A gene is emphatically not an organism. The journey from altered gene sequence to altered organism is a long and often convoluted one. It is, above all, a *developmental* journey. So now, having cast embryos aside in the present chapter, we must bring them back in the next.

CHAPTER ELEVEN

POSSIBLE CREATURES, PROBABLE CREATURES

All biologists agree that natural selection can only work when there is variation for it to use. Yet the supply of variation has been much less investigated by evolutionists than the selective process that acts upon it. The variation is not random, as is sometimes claimed. Rather, it is structured in various ways, at both genetic and embryological levels. Might this structure of the fuel for evolution be part of the explanation for increasing complexity? This question now becomes our focus of attention.

IMAGINE a world in which all male humans are as similar as identical twins, and all female humans are likewise. Also, everyone is heterosexual. Every pair-bonding is thus identical to every other. Given a few other imaginary features of the reproductive process, each brood of offspring consists of a son and a daughter who exactly resemble their father and mother.

This is a world without variation. In such a world, natural selection is impotent. It can do nothing because there are no variant humans that are any more or less fit than any others, regardless of the environmental conditions that prevail. So if the temperature rises, selection cannot favor protective darker skin; and if it falls, selection cannot favor increased blubber and a more compact shape that loses heat less readily. There would thus be no human races, just a single, homogeneous species.

Such an impoverished world seems hardly possible outside the realm of our imagination. Yet it is. We humans can create such a world. I am not talking here of the deluded desire of a small dark-haired despot to create a world consisting exclusively of tall blonds. Rather, I am talking about an equivalent world involving a different species (a tiny fruit fly). Geneticists have succeeded where dictators have failed. They have created so-called isogenic ("same genes") lines of fruit flies in which there is no genetic variation. These flies, like our hypothetically homogeneous humans, are immune to modification by natural selection, because no fruit fly is any more or less fit than any other.

Natural fruit-fly populations in the wild are not like this. Rather, they are like natural human populations in having lots of variation. And what is true of humans and fruit flies is true likewise of orangutans and praying mantises, and indeed of any other creatures you might care to mention. This is why evolution is possible.

I began with unrealistic uniformity for a specific reason. We tend to take variation for granted because it is ubiquitous. As a result, we tend not to be as curious as we ought to be about its origins. Only when confronted by an unusual variant—such as an albino—do we pause to give thought to how that variant might have arisen.

Variants can arise through two agencies: genetic and environmental. Albinos are of the former kind of causality: they result from mutation of a gene. Bronzed Norwegians returning home to Oslo or Bergen after a long vacation in the Mediterranean look different to their non-vacationing neighbors for environmental, not genetic, reasons.

Since genetic variation is inherited while its environmental equivalent is not, only the former is relevant to natural selection. Bronzed Norwegians do not give birth to bronzed babies. In contrast, albinos most certainly do give rise to albino babies, though just how predictable that is depends on whom the albino mates with and whether the particular mutation he or she possesses is genetically dominant over the normal version of the gene concerned.

I once heard an anecdote about a primitive human tribe, living in a remote corner of the world, which had an unusually high frequency of albinos. I have no idea whether it is true, but no matter, because it is instructive both about the role of inheritance in the evolutionary process and about the meaning of "fitness." An anthropologist came to study this tribe and very quickly realized why albinos were so much commoner than in other human populations. Male albinos were considered by the tribal elders to be unfit and so were denied the right to go out on hunting trips with the rest of the adult males. Left alone with the women, these albino males spent their time having sex, thus producing many albino children. In an evolutionary sense, they were very fit indeed!

So the variation upon which natural selection acts is ultimately produced by mutation, and we saw earlier that a mutation is an accidental change in a gene. But how does an altered gene, a tiny one-dimensional change at the molecular level, produce an altered creature, a much larger and often three-dimensional change at the organismic level? How can something that is so small that it is dwarfed by measurements like a hundredth of a millimeter cause the whole of a moth to turn black (as in industrial melanism), the whole of a human to turn white (as in albinism), or the whole of a snail's shell to turn from yellow to pink (another famous evolutionary case study, the details of which we can ignore)?

These questions take us from the genetic to the developmental origin of variation. Or, to put it another way, from mutation to reprogramming. That's where we're heading now. But just a couple of caveats before we go there. First, the simple correspondence between altered gene and altered appearance that characterizes black moths, albino humans, and pink snail shells is unusual; most altered appearances are underlain by changes in more than just one gene. Second, the roles of genes and environment are not always so cleanly separable as I have implied; sometimes, for example, a mutation results in an altered appearance that is manifest in some environments but not in others.

The mutant genes that evolution uses are those transmitted from generation to generation. In any particular generation, the mutant gene is present right at the start—the fertilized egg. But the egg may look no different from any other of the same species. And this is true regardless of whether we are considering large external eggs, as seen in birds' nests, or tiny internal ones like our own.

A mutant gene may fail to affect not only the appearance of the egg that contains it but also the early embryonic stages to which the egg gives rise. This is because the mutant gene is switched off. All genes have what is called an expression pattern—that is, a pattern in both space and time that characterizes their switching on and off, respectively making their (protein) product or not. For example, genes that affect eye color are expressed in eye cells, not in brain cells; and they are not expressed at all in early embryos, since these lack eyes.

Once a mutant gene gets switched on, observable consequences follow for the developing creature. Exactly when in development this switching on happens, and how pronounced the consequences are, vary from gene to gene. But in all cases we can say that the altered sequence of embryonic events results from what can be called developmental reprogramming. The original version of the gene that mutated programmed development to proceed in a certain direction; the mutant version reprograms it to proceed in another.

We have to be careful here that we do not pick up an unwanted piece of philosophical baggage. Recall from Chapter 7 ("Dances with Genes") that genes and cells interact. Looked at in one way, genes control what cells do; looked at differently, the causality is inverted, and it is cells (or particular molecules within them) that control what genes do. My use of "reprogramming" should not be taken to imply my sudden conversion to a genocentric view in which I ignore the latter part of the dance. This usage simply reflects the fact that a fertilized egg houses genes, not organs. So there are genes for eye color but no eyes. As development proceeds, eye color genes get switched on and program the development of, for example, blue eyes. In some cases, and eye color is one of them, the program is quite inflexible. In other cases, it is flexible and may be influenced by a variety of factors, including the environment. An example of this latter state of affairs is when genetically identical turtles develop as males or females depending on the temperature.

It's time to pause and reflect on where we are, where this story is going, how it connects with the chapter title, and indeed how it connects with the main theme of the book—that of evolving upward.

So far I have used particular, and often familiar, examples, such as human eye color. But now we need to begin to think more generally. Imagine a human embryo two weeks after fertilization of the egg. It is tiny. Yet it consists of thousands of cells all in a great state of flux, busy making more cells, moving around, and switching on and off multiple genes. If one or more of these genes is mutant, the embryo may develop differently from "normal." In a way, there is no such thing as normal; it is better to think of embryos as having different versions of genes and thus developing differently. I look different from my brother because our different genes caused us to develop differently; neither of us is "mutant." Though in another sense, "normal" and "mutant" are accurate, in that some mutations produce appearances that stand out from the normal range of variation—as in the case of albinos.

Anyhow, the key questions are as follows: Are some kinds of reprogramming of development possible, and others not? Of those that are possible, are some more likely to be hit upon by mutations than others? If so, might these biases in the ease of producing variants be important determinants of the direction that evolution takes, and thus of the array of creatures it produces?

Having already devoted an entire book to these questions (*Biased Embryos and Evolution*), I will be brief. The answer is yes to all of them. There are creatures that simply cannot be built through a developmental process powered by the proliferation of living cells—for example, as pointed out by the late great Stephen Jay Gould, animals with wheels rather than legs. There are also creatures that can be built from one kind of starting point but not from another. For example, earthworms have a variable number of trunk segments, and individual worms with odd and even numbers of segments are about equally common; but centipedes, which also have a variable number of trunk segments (from 15 to 191), can only make odd numbers—there are no (adult) centipedes with an even number.

So some creatures are possible, others are not. And of the possible, some are more probable than others. In tiny creatures called roundworms, it has recently been discovered that mutations more readily reduce body size than increase it.

Both qualitative and quantitative biases in the ease of producing variant creatures by mutation and developmental reprogramming may affect the direction evolution takes. This point tends to be ignored, or even sometimes dismissed, in the popular literature of evolution, some of which is disturbingly panselectionist—that is, it attributes the determination of the directions evolution takes wholly, rather than just partially, to Darwinian natural selection.

We have, at last, met up with this chapter's title, but we have not yet met up with the main theme of the book. Now it is time for this second convergence.

Notice that the variation I have been discussing so far in the present chapter is unconnected with complexity. Humans with blue and brown eyes are equally complex; so are snails with pink and yellow shells. These forms of developmental reprogramming are thus unimportant in relation to the ascents of life's ladders that are my central theme. But these ascents must also, like more mundane evolutionary events, be based on mutation of genes and consequent reprogramming of development.

The key questions themselves thus mutate. They become: (1) Are changes that increase, decrease, or have no effect on the level of complexity all possible? (2) If so, is any one of these three types of changes more probable (easier) than others? And finally: (3) Do these possibilities and probabilities contribute to evolution's tendency to produce an ever-greater average complexity of creatures?

The answer to the first question is yes. If any of the three types of variants were in some sense prohibited, evolution would have yielded a different array of creatures from the ones we see around us. Specifically, if mutations that increased complexity were impossible, all life-forms would still be bacteria. But the second and third questions are tougher nuts to crack.

Recall again our working definition of complexity: the number of different types of component parts. Recall also that parts can be cells, tissues, organs, or segments, among other things. To make a tough issue a little easier to address, I shall focus on just one of these types of parts: the cell. We have already seen that the number of cell types of which an individual creature is composed has, in some lineages, risen from one to about two hundred over the course of more than half a billion years. Is this rise due to a propensity of mutation and developmental reprogramming to produce variants with more cell types, or is it due instead to natural selection tending to favor such variants? Or is there an element of both?

Consider a simple early animal: Let's say that it is composed of one hundred cells; that all of them are the same; and that each cell contains an identical set of ten genes. Suppose this animal reproduces asexually by budding off one of its hundred cells, which, following detachment from its parent creature, divides repeatedly until the number of cells in this offspring creature also reaches one hundred. This simple life cycle repeats itself endlessly. Now suppose that in generation 1,024 the budded-off cell that founds a new creature has, for entirely accidental reasons, a mutant gene. If all the cells in its parent were identical, they must have had the same genes switched on. Since it doesn't make sense to have genes you never use, perhaps all ten genes were switched on in all cells.

But now we have a mutant form of, say, gene 8. Let's make a further assumption in this mental game and imagine that mutant-gene 8 is such that it only gets switched on when, among other things, it is on the periphery of the animal (and thus receives signals, like light, from the external environment). Because of the mutation, there are now two types of cells: skin cells in which gene 8 is switched on; and internal cells in which it is switched off. The mutation has caused an increase in complexity from one cell type to two. It has, if you like, doubled the complexity of our hypothetical animal.

Reverse mutations are also possible. Suppose that later in time, say, in generation 3,617, a mutant-gene-8 animal experiences another accidental perturbation to the stretch of its DNA that constitutes this gene. One possibility here is that the DNA sequence reverts to the one that prevailed before the earlier mutation in generation 1,024. In this case, complexity is halved. But of all the changes that could occur in that stretch of DNA, of which there are thousands, only one will re-create the original condition. Others will cause the appearance of doubly mutant versions of gene 8. Perhaps these versions will only be switched on in cells that are both external and ventral (underside), like those of a snail's foot. Such a mutation leaves complexity unchanged, as there are still two cell types; all that is different is their relative spatial arrangement.

So we have, albeit in an imaginary scenario, seen mutations that increase, decrease, and have no effect on complexity, as measured by the number of cell types in our simple animal. This reinforces the point that all three types of mutations are possible, but does it give us any insight into their relative frequency?

I believe it does. The main message, I think, is that mutations that decrease complexity at this level will be rare, as they need to precisely undo an earlier change that created a new cell type. Mutations increasing complexity will be commoner; and probably those that leave complexity unchanged commoner still.

The last of these points is irrelevant. Even if 90 percent of mutations are complexity-neutral, there will still be an upward pressure if the remaining 10 percent are biased in favor of increases and against decreases. This kind of bias will contribute to the evolution of complexity, but it is not the whole story. Natural selection is also important, perhaps even more so, because any newly complexified form has to pass the fitness test that selection sets for it. If it is unfit compared with its simpler parent, it will die out, taking its mutant gene with it.

It's time for a little caveat. In the above scenario, the starting point was a very simple one, namely a creature whose one hundred cells were all of just a single type. It could be argued that this is a special case where, to use the phrase I introduced earlier, there is "nowhere to go but up." Or, to be more precise, there is no route down. A creature with zero cell types is nonexistent. So it is hardly surprising that near the lower limit of organismic complexity, there is an inbuilt tendency for complexity to rise. What is more interesting is whether, in the long term, we can explain the tendency of evolution to produce an ever-increasing average complexity.

The problem here is that this tendency may be an illusion. We can only look backward, never forward, in terms of what creatures evolution produces. It's true that the average complexity of creatures has increased consistently thus far—providing that the milestones used to measure it are sufficiently far apart. But will it continue to do so? Will the next half-billion years produce an increased average complexity equivalent to that achieved in the last? Or, if it does not, will it at least produce an increase of some sort?

These questions lead naturally to another: Do evolutionary processes, whether here on Earth or on distant, as-yet-undiscovered planets, tend to reach some sort of plateau, where the average complexity of creatures eventually stabilizes? We can't answer this question, and will remain unable to do so unless someone invents a form of time travel. So we stick with what we know—that at least one evolutionary process, over its first few billion years, has tended toward increased average complexity.

The explanation of this increase probably has two parts, as discussed here and in the previous chapter: a bias in the origin of the variants on which natural selection acts, and a bias in how often more and less complex variants are favored by selection.

But all this is much too skeletal. Satisfactory explanations for biological (or any) phenomena need flesh as well as bones. So it is time to get more specific. And one of the most important specific processes involved in the ascent of life's ladders is embodied in the phrase "duplicate and diversify." So this, a process championed by the Japanese geneticist Susumu Ohno, is our next port of call.

CHAPTER TWELVE

DUPLICATE AND DIVERSIFY

The diversification of duplicated, or more generally replicated, parts of creatures is probably the single most important causal process in evolution's production of rising complexity. So this chapter is, in a sense, the book's heart. Well, actually, I should include the next chapter, too (a second heart chamber perhaps?), as that continues the same story. The splitting of the story works as follows: here I discuss the general principle of duplication and divergence, and give some examples of it at the level of big parts of organisms, such as segments in animals and leaves in plants; in Chapter 13, the focus shifts down to the level of genes.

OUR central story begins with a key feature of the evolutionary process whose effect on the way the process works cannot be overstated. This feature is the unbreakability of the flow of life across the generations. Breaks in evolution, in an overall sense, are impossible. In an evolutionary lineage, even a break of a single lifetime in a great chain of ancestors and descendants spells the end of the line: extinction.

What this inescapable fact means is that evolution cannot afford the luxury of a design process in which temporary bad designs act as a conduit to subsequent good ones. Contrast this situation with the one of a human designer working for a car manufacturing company. He/she can play around with a series of designs, including some that a colleague might point out are fundamentally flawed in some way, yet ultimately arrive at a wonderful one that leads to the production of what becomes a bestselling car. Those flawed designs that may have aided the evolution of the designer's thought end up in the office waste bin. This experimentation is not a problem, as the old model of the car concerned is unaffected. People continue to drive around in the old model, blissfully ignorant of the fact that a possible successor to it would have been a disaster.

In evolution, the design and production processes are inseparable. Design does not happen on a drawing board or in a computer—it takes place by tinkering with the "car" (well, creature, actually) as it goes about its business.

This fact does not mean that evolution cannot do experiments. Actually, it is doing them all the time, as mutations in genes keep on happening, albeit at a very low frequency per generation, just by accident. Every mutation is subject to that ultimate test of natural selection: if it renders its bearers inviable or infertile, it disappears. Such problems are not terminal, because most species, most of the time, are represented by millions of individuals, and most of them are expendable, in that their deaths at pre-reproductive ages are not problematic for the continuation of the species.

Indeed, many species have huge overproduction in every generation. Picture a mass of frog spawn in a pond. There may be several thousand eggs in a batch produced by a single female. Most will die as tadpoles, before they get the chance to metamorphose into adults. In the callous world of nature, this is just the way things are. And a few eggs or tadpoles may die not through being eaten by a vicious dragonfly larva but rather because they are the bearers of a mutant gene that somehow causes the arrest of their developmental process.

So how do we square up these apparently incompatible notions:

that evolution in some sense cannot afford the luxury of failed designs, and yet in another sense is experimenting with them all the time? Is this not a contradiction?

The reason it is not a contradiction goes something like this: Mutation has a luxury that natural selection does not. A mutation affects an individual, and individuals are, as we have seen, expendable. But natural selection can spread a mutation through an entire species. Suppose it does just that with a particular mutant gene that is advantageous in the environmental conditions that prevail at the time but lethal under other conditions that might follow later. This possibility illustrates the blindness of natural selection, which was made famous by the English biologist Richard Dawkins in his book *The Blind Watchmaker*.

We have no way of telling how many lineages have met their doom because of a long-term loss that was an inevitable but unseen (by selection) corollary of short-term gain. There have probably been lots of them.

Now, all this has a very important connection with the complexity of creatures, measured in terms of their number of different types of component parts. Probabilistically, losing a part is a riskier venture than gaining one. Mutations that alter development so that parts get lost are less likely to survive than those that alter development so that additional parts appear. In the latter case, the extra parts are likely, in the first instance, to be straightforward copies of ones that already exist, as when a centipede adds extra segments, and the main risk is that of redundancy, which can be thought of as merely a nuisance factor. But in the case of losing a part, the main risk is the much more serious one of losing some vital function. For example, if a fruit fly loses its middle thoracic segment, it may perhaps still be able to walk—on four legs rather than six—but it will certainly not be able to fly, as it has lost the only segment that has wings.

The lineages most famous for sliding down life's ladders rather than climbing up them are parasites. If, like a tapeworm, you live in the plentiful food supply of another creature's gut, you can afford to lose various things that free-living animals cannot, such as eyes. In the nonparasitic world, eyes (and other parts) are sometimes lost for other reasons. For example, fish that live in the lightless waters of caves are often characterized by the reduction, or even loss, of their eyes, though this may be offset to some extent by the enhancement of other sense organs.

Lineages that lose parts will often end up as evolutionary cul-de-sacs, whereas those that add parts may go on to greater things. So how does this happen? I'm going to deal with the two sides of the coin separately: first duplication/replication; then divergence/diversification.

The phrase "duplication and divergence" has a certain alliterative appeal, which probably accounts for its commonness in the evolutionary literature, even though it is merely a special case of replication and diversification. In any event, whether we are thinking about processes where something becomes duplicated or more generally multiplied up in number, the end result is that a creature comes to possess more than a single copy of a part.

Let's focus for a while on arthropod segments. These structures begin to be formed during development as transverse stripes of tissue running across the embryo in which particular genes are switched on and so are making their protein products. These proteins can be thought of as sending signals to the overall developmental process, such as "make a segment boundary here." If something causes more such stripes to be formed than usual in a certain individual, that individual will, later in its development, have more segments than usual. The "something" might be a mutation in another gene that controls the positioning of the stripes of the one we started with; or it might instead be an environmental influence, such as increased temperature, that somehow deflects development from its normal course.

Whatever the segment number of the first-ever arthropod, it is certain that both increases and decreases in number have occurred in its various descendant lineages. The same may be said of the number of vertebrae (and associated ribs and muscles) in vertebrate animals such as ourselves. And in plants, the number of leaves has undergone changes in upward and downward directions, though this example is messier because there is lots of variation even among different individuals of the same species. Humans all (or almost all) have the same number of vertebrae. But maple trees certainly do not all have the same number of leaves.

What segmented animals and trees have in common is a modular design. Each of the many modules in a given individual—whether segments, leaves, or whatever—has a common developmental program, not so much in the strict sense of being identical, but rather in the looser sense of being similar (to a variable degree).

Now, if you find yourself, at any point in evolutionary time, as a modular organism that has lots of more-or-less-identical modules all doing the same thing, and you do not need as many as you have got in order to do that thing adequately, you have redundancy, which can be exploited by evolution. New types of parts can be created by diversifying some of the existing modules to do new things while still leaving enough of the old ones to do whatever it was they did in the first place.

The process of diversification requires no special magic. In fact, we already looked at how it might work, at the cellular level, in the last chapter: starting with a single type of cell, a mutation that causes cells to develop differently in response to light will lead to the appearance of skin cells and internal cells. And although segments and leaves are much more complex than cells—indeed each consists of thousands of cells—the principle of diversification of overrepresented modules is entirely similar at these different levels of organization.

It's time for some specific examples. We'll look at the following, in sequence: the front end of vertebrates, the front end of arthropods, the back end of arthropods, the flowers of higher plants, and the brain of an octopus.

Like all the major groups of animals, vertebrates started off in the sea. They thus had to breathe water, not air, and so had gills rather than lungs, just as do the fish of today. But unlike most of today's fish, the earliest vertebrates had no jaws with which to capture their prey. Instead, they had a primitive kind of mouth that took the form of a roundish hole, and they sucked up detritus through this opening, digested whatever was digestible, and ejected what was not in their feces. They were a kind of living vacuum cleaner. Today a few species still have this design, but it has largely been replaced by the more effective design that involves upper and lower jaws.

So at a particular point in evolutionary history, vertebrate jaws appeared as a novel feature. But where did this novelty come from? To provide a possible answer to that question, we need to shift our attention from feeding to breathing, and so from jaws back to gills. Vertebrate gills consist of a series of arches of tissue on each side of the head, separated by slits through which water passes. Each gill has an abundant supply of blood vessels, just as do our own lungs.

As water flows through the gill slits, gaseous exchange occurs between the water and the blood. That is how a fish acquires its oxygen and loses its carbon dioxide.

Those primitive vertebrates that had jawless mouths had multiple gills. Each gill was a module, in the same sense as segments and leaves are modules. And since there were several on each side, there was redundancy, especially given that if fewer were to do the same job of breathing, they could easily be made bigger, size increases being one of the easiest things for evolution to achieve.

So we have replicate modules and redundancy; now divergence of structure and function enters the picture. Each gill arch has a skeletal support, be it cartilage or bone. Now, here is an unusual thing—evolution thinking out loud. Suppose, it says to itself, I modify the anteriormost arch or two into jaws, leaving enough posterior arches to do the job of breathing; it might just work. In reality, thinking out loud takes the form of the natural experiment provided by variation among individuals, its ultimate source being in mutations of the genes that control gill arch development.

The only glitch in this otherwise straightforward story is that thorny old issue of the usefulness of intermediate structures—in this case those that are halfway between gill arches and jaws and might not be much use for breathing or biting. There are three possible ways out of this apparent impasse. In a few cases, it may be that intermediate structures smoothly pass from one function to another, through a phase when they get by doing a bit of both. In other cases, a new structure may begin to evolve for one reason and only later continue to evolve for another. One theory of the origin of birds' wings is that the first feathered forelimbs were used for trapping prey, and only later, after further elaboration, for flying.

In other cases still, there may be no intermediate stage, with development being deflected suddenly from one course to another.

Although such "saltationist" theories of evolution, when advanced at various points in the last century by several biologists, did not gain acceptance, and probably rightly so, as *general* mechanisms of evolution, some specific examples seem to cry out for this kind of explanation, including the origin of the turtle's shell. A turtle has its shoulder and hip girdles internal to its shell. Hence the shoulder girdle is internal to its ribs, as these are fused to the shell. How can you have a girdle that is halfway in and halfway out? Clearly you can't. Mutations that cause sudden big changes in animal development are not the problem here, as many of them are well known. Perhaps the most famous is the one that makes an extra pair of legs grow out of a fly's head where its antennae should be. Rather, the issue is that such mutations usually lead to decreased, not increased, fitness. But in the vastness of evolutionary time, a fraction of one percent of big changes in development being fitness-enhancing is all that is required for the origins of very occasional novelties by this mechanism.

And now, from vertebrates to arthropods. If you look carefully at an arthropod feeding, you will notice pairs of little appendages on the underside of the head that the animal uses to manipulate its food. The exact number and nature of these appendages vary among the different arthropod groups. So exactly what you see will depend on whether you are observing the feeding behavior of a locust, a spider, a centipede, or a lobster.

Regardless of their precise form, where did these feeding appendages come from? The answer is as before: diversification of replicated parts. The earliest arthropods had many similar segments, most of which bore a pair of "legs." These structures were not used for walking on land, since the creatures concerned were marine, so thinking about the legs of a shrimp will give you a better mental picture than thinking about the legs of a butterfly.

Although many legs may make light work of getting about, if there are enough of them it will be no great loss if some become specialized for other purposes, like feeding. So modification of the structure of the anteriormost pairs is not a problem, and this is indeed what has happened in arthropod evolution. Thus the story is much like the vertebrate gill-to-jaw one in its essentials, albeit it differs in detail.

Now for another shift in focus, but this time not from one major animal group to another but rather from anterior to posterior. This story involves a particular group of arthropods, namely centipedes.

If you look at the rear end of a centipede, what will you see? Well, the chances are that you do not yet know the answer to this question, because, like most people, you have never paid much attention to the rear ends of centipedes, or even centipedes generally. But I can tell you that you would see a strange-looking pair of legs that are not used for walking. They are longer than normal legs, often thicker, too, and they project backward, not sideways. They point slightly upward and rarely make contact with the ground.

The function of these odd appendages is not yet entirely clear. In fact, they probably have several functions. Since most kinds of centipedes can move backward as well as forward, the rear legs most likely have a sensory function—a bit like a spare pair of antennae kept at the back for probing the lay of the land when retreating. But they probably have a sexual function, too, since in many species they are quite different between males and females (usually fatter in males), in contrast to ordinary legs, which are the same in the two sexes.

Even if you are one of those minimalist centipedes with "only" fifteen leg-pairs, you can certainly manage to walk or run using fewer than this number. So the divergence of one of the leg-pair modules to a different function has no adverse effect on getting about, but it does have a positive effect on other things. Again, this is a classic case of the diversification of replicated structures.

And now, a rare foray into the world of plants. I have concentrated more on animals throughout this book because I am a zoologist, not a botanist, so I know more about them. The following will at least go a small way toward redressing the balance.

A little earlier I mentioned a maple tree. Almost everyone can recognize this tree species, and anyone who is uncertain about what a maple leaf looks like has only to inspect the Canadian flag. In my own case, I grew up in a house with a huge maple tree in the front garden, and I recall often looking up into its impressive foliage, especially in autumn, when it was a kind of bright natural gold. As a child, I admired it, and often wished it had lower branches so that I could climb it. Only much later, as a biologist, did I ask myself questions like: How many leaves has it? I still don't know the answer. But this is one of those questions to which only a rough answer is needed, such as "thousands."

Maple trees are members of the most complex group of plants—those with flowers. Maple flowers are not as conspicuous as those of some other types of trees, such as the horse chestnut. The size, shape, and color of flowers vary immensely, as everyone knows. Venture beyond the realm of trees and we find the very different flowers of daffodils, grasses, orchids, daisies, and roses. Flowers have inspired designers for centuries. But where did they come from?

Many simple plants do not have flowers. These include mosses and ferns. Most people know that the coal we burn in our fires is the fossilized remains of ancient forests. But what is not so widely known is that those forests were very unlike the forests of today. They contained no maples, chestnuts, or oaks. They were made up largely of tree ferns.

Maples and their cousins were absent from the coal-age forests because these forests were alive about 300 million years ago. The first flowering plants did not appear until many millions of years later—perhaps about halfway between then and now.

Just as vertebrates with jaws evolved from their primitive jawless counterparts, so flowering plants originated from flowerless ones. And they did so by exactly the same evolutionary trick: diversification of replicated parts. Every leaf is a module. The main function of a typical leaf is to photosynthesize—that is, to turn the inorganic energy of sunlight into the organic energy that powers not just the plant but ultimately the entire ecosystem, including animals like ourselves.

Just as a centipede can afford to diversify some of its many legs for purposes other than walking, so a plant can afford to diversify some of its many leaves for purposes other than photosynthesizing. The petals of a flower—and indeed other, less conspicuous parts of flowers, too—are actually modified leaves. They are versions of that same developmental module from which leaves are made, but in the case of petals the program of development has become altered so that it delivers a visually and functionally different end product. Again, the ultimate source of the variation involved is mutation, and the spread of a mutation from the single individual in which it first arose to the whole species is brought about by Darwinian natural selection.

These stories of the origins of evolutionary novelties such as jaws and flowers are fascinating. But how do we know that they are true? So far, I have skirted around this issue, so in a way the stories as they have unfolded thus far take the form of mere assertions. We must proceed beyond this level of explanation to look at the sort of evidence that has led biologists to believe in the importance of the diversification of replicated modules.

One important kind of evidence in favor of two different structures having had a common evolutionary origin is that they have major similarities, particularly in the way their component parts are ordered, despite their many differences. Human and ape arms both have a pattern of 1-2-many-5 if we count the bones of the upper arm, the lower arm, the wrist, and the bases of the fingers. This points to a shared origin. More important for our story about duplication and divergence, human arms and legs also share this pattern. This does not mean, of course, that we evolved from armless ancestors who ran around and attacked their prey by head-butting. But it may mean that in the more distant evolutionary past, the two pairs of fish fins that evolved into legs when vertebrates first invaded the land had their origin in a single pair, which then duplicated.

Other kinds of evidence, in addition to shared structural components, also point to a common evolutionary origin of structures that now look quite different. Similarities in embryological development collectively constitute an important pointer. So, too, do similarities in genetic control, as we will see in the next chapter. The best approach is not to rely on any one line of evidence for a common origin, but rather to look at all the available evidence and use that as your guide. The result is that in many examples, including most of those we have been discussing, the conclusion that different structures have arisen by divergence from replicated modules is overwhelming. In other examples the picture is less clear, but it will doubtless clarify with time as more information on the development and genetics of the structures concerned becomes available.

There is one more example that we have yet to think about before we head off from the structural realm to its genetic counterpart: the brain of "the octopus." Actually, there are about two hundred different species of octopuses in the world, but this need not concern us. All are brainy by comparison with their many molluskan cousins, such as slugs, snails, and oysters.

There is an important difference between this example and the other animal examples I have discussed, involving arthropods and vertebrates. These two great groups of animals are based on a segmental body plan. The mollusks, in contrast, are not. There are no equivalents, in a mollusk, of centipede segments or human vertebrae. It's true that occasional mollusks may have a few features that are reiterated in a quasi-segmental way, such as the series of shell plates in those strange mollusks called chitons. But in most mollusks the shell is a singular structure (snails, for example) or a double one with left and right valves (oysters and such).

It would seem, then, that the molluskan design is less modular than its vertebrate or arthropod counterparts, and thus that the whole idea of diversification of replicated parts is less applicable. But, on the contrary, the idea is indeed still applicable; it just needs to be dealt with at a different level.

The brains of slugs and snails are simple affairs. They consist of several small groups of nerve cells connected together. In the octopus, the main advance on this basic design has been to scale it up. The brain of an octopus is much bigger than that of a snail, even if we allow for body-size effects. For example, we might compare the brain of a small octopus with that of a similarly sized giant African land snail (the ones famous for being partial to beer). The small octopus's brain is very much larger than that of the giant snail.

This increase in size is achieved, for the most part, not by having more physically distinct groups of nerve cells (ganglia) but rather by scaling up the size of each group, which is in turn achieved by increasing its cell number. So the level at which the divergence of replicated parts occurs is that of the cell. The big brains of octopuses not only have more cells than their sluggish relatives; they also have more types of cells, and also more connections (and types of connections) between those cells. This increased complexity at the cellular level enables octopuses to perform behavioral feats that are quite beyond the capability of most other mollusks.

So, having completed our grand tour of specific examples, let us now return to the point at which we started this chapter. There is no harm in restating this general point: diversification of replicated parts is the single most important evolutionary process leading to increased complexity. It has been at work in many different groups of creatures, and it operates at several different levels. We have seen its operation at the levels of organs, segments, and cells. Now we turn to yet another level at which it operates: genes.

CHAPTER THIRTEEN

THE BEST BOX OF ALL

Although genes can be tough going for the general reader, our story of the rise of animal complexity would be seriously incomplete without a foray into the genetic realm. In fact, we already entered this realm in Chapter 7 to see how complexity increased during embryological development. Now we return to it in order to see how, in relation to evolutionary increases in complexity, the duplication-and-divergence principle of the previous chapter applies not just to things like segments and leaves but also to the genes that make them.

ONCE upon a time geneticists were fascinated by crosses; nowadays they are more fascinated by boxes. The founder of genetics, the nineteenth-century monk Gregor Mendel, spent his time in the monastery garden crossing tall pea plants with short ones, or those that had smooth peas with those that had wrinkled ones, and examining what the offspring pea plants looked like in the two succeeding generations. By doing these simple crosses, he worked out how the features not just of pea plants but of all plants, and of all animals, too, are inherited from one generation to the next. A beautiful general theory emerged from his simple experiments, and it has stood the test of time.

Crosses were important in the early days of genetics, not just in the nineteenth century but a fair way into the twentieth (when the favorite workhorse on which to do them shifted from peas to flies), because the primary problem that genetics as a science set out to solve was that of inheritance, and crosses were an essential tool in investigating this problem. But once it was solved, the focus of genetics shifted. The things that carry information about how to build a creature, and how to make its various features, were, according to Mendel, rather anonymous and invisible things simply called "factors." Later they acquired the more specific name of genes. In 1953, they became visualizable, in the sense that Watson and Crick discovered the structure of the molecule of which they are made, namely DNA, albeit they were still invisible, at least to the naked eye.

Clearly, then, one shift of focus was from the large-scale features of animals and plants that are the results of gene activity to the genes themselves, small-scale entities at the molecular level of organization. But this shift of focus is not the only important one that has taken place in genetics. Another important shift, made possible by the organism-to-gene one, was from genes as bearers of hereditary information across the generations to genes as things that produce the proteins that build and run an organism and the cells of which it is made. This shift, paradoxically but pleasantly, takes us back from gene to organism. One of the most exciting areas of genetics today is developmental genetics, in which we study how genes affect the developmental process that takes an organism from egg to embryo to adult. We have, therefore, come full circle: the organism was the focus of attention in Mendel's day; the gene was focal in Watson and Crick's day; the organism is again a major focus of attention now.

But in going round in this quasi-circle, we have learned a lot about creatures. We do not have to treat their features as inscrutable anymore; rather, we can investigate how they arise as the creature concerned develops.

So we are now in the era of molecular and developmental genetics, and this is where boxes come in. As many genes have been studied in many creatures, we have come to notice that certain DNA sequences keep showing up all over the place: sometimes in different genes in the same creature; sometimes in the same gene (that is, one that makes the same protein) in different creatures.

These sequences involve what has sometimes been called the genetic alphabet. Like the text you are reading right now, the DNA of which genes are made is a unidimensional thing—essentially a long string of characters. In text, there are twenty-six characters that we call letters, along with several others concerned with punctuation, such as commas and periods. A sentence—think of it as equivalent to a gene—is made up of a long series of characters. The one I have just finished has about ninety of them.

DNA has only four kinds of characters—building blocks that can be given their unfamiliar and hard-to-remember chemical names or, better, simply be represented by the initial letters of those names—A, T, C, and G. A typical gene is made up of many more characters than a typical sentence. Instead of ninety characters, it might have nine hundred, or even nine thousand. Imagine looking at the exact sequence of part of a gene—say, the first ten characters. Suppose it reads AAATTTCCCG. Now pick another gene at random—one that makes a completely different protein. Its first ten characters might be TTAACCGGTT. Clearly, these are very different sequences. There is one accidental coincidence, an A in position 3, but otherwise they have nothing in common.

If you roam at random through the DNA of a human, a fly, a

pea plant, or any other sort of creature, and compare two sequences of the same length, this is typically what you will find—they will be very different from each other. But occasionally they are not, and this is where things get really interesting, and also where boxes come in.

In the early days of molecular genetics, it was discovered that many genes have the short sequence TATA just before their start. To highlight this common sequence in the otherwise different streams of characters, it became accepted practice to draw a box, or rectangle, around it. This practice is helpful, because long strings of characters tend to befuddle the eye—they bury it in too much information and prevent important recurring patterns from leaping out. And so we had the TATA box. Since then, many other commonly recurring sequences of varying lengths have been found, and been called boxes, too. The way they are named varies. Sometimes the name derives from the repeated sequence—as in TATA—but sometimes the box is named after its discoverer. A third approach to naming a box is to base the name on something the repeated sequence does, when it is eventually translated into its protein product. This is the rationale behind the name of my "best box of all," the homeobox.

To understand this box's name, we need to travel back in time again, not quite so far as Mendel's time (he published his results in the 1860s), but to the closing decade of the nineteenth century, the 1890s, and the year 1894 in particular. This was when the English biologist William Bateson wrote a monster of a book with the uninspiring title *Materials for the Study of Variation*. Bateson was fascinated by characters that varied in a discrete way, such as the number of segments in an arthropod, which is always an integer, rather than in a continuous way, like human height, where people are not restricted to integer heights like 5 feet or 6 feet, but rather can have any of a continuous series of values—for example, 5 feet 9.768442 inches.

One thing Bateson noticed in his studies of discrete variation in general, and of arthropod segments in particular, was that occasionally there were individual creatures that, although they had the right number of segments for their species, had one segment of the wrong form. But not just wrong in a general sense; rather, wrong in the specific sense that the aberrant segment possessed structures that were normally found in other segments. The examples that have become most famous, largely through the subsequent work of the American geneticist Ed Lewis in the mid-twentieth century, are flies in which legs grow out of the head where antennae would normally grow, and flies in which the little pair of flight-balancing organs ("drumsticks") that normally grow out of the last thoracic segment have been replaced by an extra pair of wings.

Bateson called the phenomenon of having the right thing in the wrong place homeosis. As usual, the prefix "hom-" refers to being the same, as in "homogeneous" and "homosexual." So, for example, the fly with legs growing out of its head has appendages that are the same (approximately, anyhow) as the legs growing out of its thorax. Later, when geneticists had coined the term "mutation" for an error in the structure of a gene, the phrase "homeotic mutation" was used to describe the particular class of mutation that gave rise to homeotic structures.

And so to the homeobox. In the early 1980s, two groups of biologists, one working in Indiana, the other in Switzerland, made an important discovery. Actually, that is putting it mildly: if I had to choose the most important biological discovery of the last half century, this would be it. They discovered that a certain DNA sequence kept showing up in a whole lot of genes in many different animals.

This box was much longer than the TATA box. In fact, it consists of about 180 characters of the DNA alphabet. And although the sequences vary a little, especially across big evolutionary distances, such as that separating mice and flies, the degree of similarity of the sequences is staggering—often more than 90 percent; sometimes more like 99 percent.

At first, the homeobox was thought to have something to do with segmentation, as all the animals initially looked at were segmented. And it is true that all those genes that cause homeotic segmental effects when they mutate do indeed contain homeoboxes—hence the name. But in the end, further studies of a wider range of animals showed that all animals had homeobox genes, and indeed lots of them. So the proposed specific link with segmentation turned out to be wrong, but a more general link with animal development, as we will see, turned out to be right.

I have managed to get this far without explaining much about how genes work. But to progress from our present point of understanding how widespread homeobox genes are to understanding why they are so important, we have to get into this issue at least a little.

Recall that a typical gene is a sequence of hundreds or thousands of DNA characters. The way it works is to make a protein. Different genes make different proteins, and different proteins do different jobs. Your hair is made of a protein called keratin; your red blood cells are packed with a protein called hemoglobin; your metabolism is powered by many proteins all in the general category of enzymes; and your development as an embryo is driven by many other proteins, including a very important group that we'll call "gene switchers" (transcription factors, for those in the trade).

Here we encounter again the crucial point made earlier in the book—especially in the chapter "Dances with Genes"—that the control of embryonic development can be looked at in two ways, both of them equally valid. From one point of view, genes are in control; from another point of view, control is in the hands of proteins. But both of these viewpoints are oversimplifications of the real situation of an interactive "dance" between the two types of players. Genes make proteins, but the proteins that some genes make control the activities of other genes. They do this both qualitatively (switching on and off) and quantitatively (fine-tuning the rate of activity of the gene concerned).

What proteins do and where they must go are connected. Hemoglobin must wander the body to undertake its role as a gas exchange agent to all the tissues. Gene-switcher proteins, on the other hand, must go into the central nuclei of cells where the genes reside, and attach themselves to the DNA in order to exert their switching effects. This attachment requires a particular structure, just as the USB storage devices I'm using to store backup versions of this chapter require a particular structure to connect with the USB port on the side of my laptop.

The homeobox provides that structure. The 180-character homeobox of a developmental gene makes a region of the corresponding protein called the homeodomain. This domain anchors the protein to the DNA of those other genes that it will act as a switching agent for.

Of course, different switchers operate on different target genes. So they must not be entirely the same as each other. But this requirement for some variation does not present a problem, because just as the homeobox is only a small fraction of the overall sequence of a homeobox gene (180 DNA characters out of 5,000, for example), so the homeodomain is only a small fraction of the corresponding protein. The common domain enables anchorage, but the other domains, which differ from one switcher protein to another, give specificity of action.

Now, the whole point of this foray into the territory of the homeobox is to link the realm of genes (and especially those that control development) into our story of the evolutionary importance of diversification of replicated parts. In the previous chapter, as you will recall, this story first unraveled as a story about bigger-scale structures such as gill arches, segments, and leaves.

Every time a cell divides, the genes get copied so that a complete set goes to each daughter cell. This is true of almost all cells, wherever in the body they are found. Included in this, and of special importance, are the cells of the germ line that lead to the production of the sperm and egg cells that will create the individuals of the next generation—albeit in some of these cells a kind of cell division somewhat different from the usual one occurs.

Every time genes are copied, there is a risk of making mistakes, just the same as every time I type the same word. In the course of writing this book, I have frequently had to alter "developmenyal" to "developmental." Also, occasionally, I have typed the same word word twice in tandem, just like that. The first of these typographical errors is like an ordinary gene mutation; the second is like a rather special kind of mutation—a duplication.

Genes can get duplicated in two main ways, both of them entirely accidental. One of them involves a localized process where, for example, a DNA strand slips a bit and a single gene gets duplicated. The other involves bigger-scale processes such as the accidental duplication of the entire genetic material. Both kinds of accidents have happened many times in evolution, and both have happened many times in the evolution of the homeobox genes.

Any one animal has hundreds of homeobox genes. They constitute a sort of family. They show different degrees of relationship. Some are similar enough to look like siblings. Others, a bit more distinct from each other, look like distant cousins. The reason for this pattern is that long ago in what we might call a stem creature—the common ancestor of us all—a single ancestral homeobox gene duplicated, and the redundant second copy evolved a new function. Later, the same thing happened again. Then it happened again and again over hundreds of millions of years.

This process is the single most important one in the elaboration of the genetic material of creatures over evolutionary time, in the same way that the diversification of replicated structures has been the most important cause of increasing complexity of the body. In both cases, simpler forms have remained and prospered rather than being replaced by their more complex cousins. Bacteria still outnumber us animals in today's world, testifying that complexity of genomes and/or bodies is by no means a prerequisite for ecological success. As it is important to keep emphasizing, what has increased over evolutionary time is the maximum complexity of creatures, not the minimum—which has remained more or less the same for eons.

Now an interesting question arises: Does the process of diversification of replicated parts run in parallel at the two levels—the genetic and the structural ones? Do creatures with complex structures also have complex genomes? The answer, it seems, is that famous apparently self-contradictory one of yes and no—though I think that the yes bit is the more interesting half of this duo.

I have told the story of duplication and divergence of genes using the homeobox family as my case study because, in terms of building bodies, this family is especially important. But other families are involved, too, such as the globin family, to which hemoglobin belongs. In fact, most of the genome is composed of more or less extensive gene families, all having arisen through repeated duplication, or replication, of their founding father. On an *a priori* basis, it would seem that there is a good reason to expect a complex family of developmental genes to go hand in hand with a complex developmental process leading to a complex body. In the case of other gene families, those not involved in development, it is less clear what we should expect. So let's restrict our attention to the former case.

It is generally true that more complex animals have more complex families of homeobox genes. But no evolutionary stories are simple, and I want to mention two things in particular that urge caution in making too clean a connection between the diversification of replicated genes and that of replicated body parts.

For both of these caveats I will focus on a specific subgroup of homeobox genes that are involved in the patterning of the main body axis in all bilaterally symmetrical animals, whether worms, flies, fish, or people. The genes of this subgroup go by an abbreviation of the word "homeobox"—they are called the Hox genes. I oscillate between thinking that this is a helpful label and thinking that it is unhelpful because some people will assume that the two categories are the same rather than that one is a subset of the other. But in any event, the Hox label has taken hold, so we are stuck with it.

The simplest animals, such as sponges, lack bilateral symmetry (or any symmetry at all), and correspondingly lack Hox genes. Basic bilaterians like flatworms and roundworms have Hox genes, but not very many of them. More complex invertebrates, like arthropods, have more; and the most complex of all animals, the vertebrates, have many more still.

Now for the first caveat: the correspondence is far from perfect. Most fish, for example, have more Hox genes than we do, even though they would generally be considered simpler creatures—certainly with simpler brains and so simpler behavior patterns.

The second caveat involves the causality of increased bodily complexity in terms, as usual, of the number of different types of component parts. A butterfly is more complex than a common garden centipede because, instead of having many similar segments, it has specialized different segments in different ways: some have wings, some have legs; two have both; several (in the abdomen) have neither. But a butterfly does not achieve its greater segmental complexity by having more Hox genes; rather, the enhanced segmental complexity derives from a different way of deploying essentially the same set of genes.

So the diversification of replicated parts is a key process—perhaps *the* key process—in the ascent of life's ladders in both genetic and structural domains. The relationship between the two domains is messy, but that is to be expected of a process whose very nature is the imperfect sifting of a series of accidents in an ever-changing environmental milieu. The wonder is not that relationships are messy but rather that they can be discerned at all.

Science is a constant interplay between the particular and the general. Too much focus on the former leads to a kind of bookkeeping mentality; too much on the latter can lead to grand theories that are entirely without substance. This chapter and the last have included plenty of particular examples, but their focus has been on a single underlying principle of broad applicability across the whole living world. In the next few chapters, the focus is inverted: they include references to general principles, but they concentrate on particular evolutionary events—particular steps up life's ladders. They start with the origin of animals and end with the origin of book-reading humans.

CHAPTER FOURTEEN

FROM SIMPLE TO COMPLEX

The simplest animals can be found in the sea. The jellyfish that we carefully avoid while swimming, the sea anemones that live in rock pools, and the sponges that can be seen growing on the seabed by an observant snorkeler are all creatures that have retained their simple body form more or less unchanged in its essentials since the dawn of animals, hundreds of millions of years ago. But "simplicity" and "complexity" are relative terms. How did these simple-yet-complex animals arise from simpler unicellular ancestors?

THE first upward step in organismic complexity about 3.5 billion years ago was not made by animals, because these had still to wait a very long time to make their appearance. Rather, it was made by microbes. Indeed, the first step of all could be said to have been made by something that was more a chemical than a creature. It is a philosophical point as to when an aggregation of large molecules can be said to have moved from one realm to another: the nonliving to the living. Perhaps in structural terms the acquisition of an outer membrane, and in functional terms the ability to reproduce with tolerable reliability, were the key features.

Anyhow, if we regard the origin of cellular life as step 1, and the origin of complex (eukaryote) cells as step 2, then the origin of multicellular creatures was step 3. I'm going to start here, three rungs up from the mud, to tell my simple-to-complex story.

This is an appropriate point at which to remind ourselves of the nonlinear nature of life—or, to be more specific, of the existence of several ladders that creatures can climb to rise above the unicellular lawn. At the very least, there have been three: the animal, plant, and fungal ladders. I'm going to concentrate on the first of these, as I have been doing throughout this book.

So what do we know about the origin of animals? What kinds of creatures were the ancestors of animals, and what did the first animals look like?

The world is awash with creatures that consist of single cells. If they were all scaled up in size by a few orders of magnitude so that we could see them, we would be truly amazed. They have a bewildering variety of names, some familiar, some not: bacteria, archaea, yeasts, amoebae, flagellates, ciliates, and many more. The last three of these used to be collectively called protozoans, or "first animals."

Although this name has recently fallen from favor, because we restrict "animals" to multicellular forms, it acts as a sort of pointer in our inquiry into animal origins. Of all the diverse groups of unicells, it is probably among the protozoans that the creature that spawned the animal kingdom is to be found. Well, not "is," I suppose, but "was," because we must always remember that evolution never stands still. Some present-day protozoans may have changed little from the elusive ancient one that was our ancestor, but they won't be exactly the same.

Much time and effort have been put into trying to decide which particular subgroup of protozoans (flagellates, perhaps?) looks most likely to have included the animal ancestor. But I'm not going to pursue the issue in this way. Rather, I want to look at what features the first animals had, and how they came to acquire them.

The most important thing, of course, was becoming multicellular. So we should ask: What is the most crucial feature allowing the construction of a multicellular body? The answer is "sticky stuff," whether you choose to call it cement, glue, or Velcro. A typical protozoan cell doesn't stick to other such cells. Rather, each cell is an island; or, to be more exact, each cell is an organism. In contrast, you and I are full of sticky stuff holding our cells together. If it suddenly disappeared, we would collapse from several-feet-tall normality and end up as a several-feet-wide stain on the carpet. It took us many millions of years to climb life's ladders; without intercellular Velcro, it would take us only minutes to undo that long climb.

So the nature of this living Velcro is worthy of some attention. Let's start with its location. This is predictable enough: if its job is to make cells stick together, then it must be a feature of their surfaces rather than of their interiors. And so it is. All animal (and other) cells are enclosed in a thin membrane composed of two things: proteins and fats. (So much for a fat-free diet—not a good idea if taken literally!) Over much of the periphery of an animal cell, this thin membrane is uncomplicated by anything else. But if you were to zoom in on a particular area of membrane, and look at it in more detail, you would see that, here and there, it is peppered with other things—things that stick out untidily from its otherwise smooth surface.

Among these protrusions are molecules with sticky ends that, when they find another sticky end, on a nearby cell, grab hold of it and, well, stick. How effective this process is depends on, among other things, the shape of the cells concerned. Some cells are quasi-cuboid, in which case the large flat surfaces afford great sticking power. Others are more nearly spherical, and these are less good at sticking together.

In a complex animal like a human, there are all sorts of complications, including the existence of many different types of sticky molecules. Often, one of these is found in a particular cell type and prefers to join hands with others of the same type: this is why blocks of the same tissue retain their coherence. We take it for granted that there are no muscle cells interspersed with nerve cells in our brains. One reason for the lack of such interspersion is that brain cells and muscle cells have different types of Velcro.

But the first animals had no such problems to solve. They probably had only one cell type, and thus no cell-type-recognition requirement of their sticky molecules. So they just needed one type of these molecules for general adhesion purposes.

But where did that one type come from? How does a creature go from no sticky transmembrane molecules to one? The answer, almost certainly, lies in the process of duplication and divergence that we examined in the last two chapters. If a gene producing a nonsticky protein molecule with some other function duplicates, then one of its copies is redundant. Following mutations in this, it will change in unpredictable directions. Some mutations may make it utterly useless—for example, turning it into a protein that is so short it can barely have any function at all. But others will, by accident, change it in other ways. And one such way is to become sticky-ended.

There's more to being a multicellular creature, even a minimalist one, than having your cells stick together. Your shape—or body form—is important, too. All animal cells need to breathe in the sense that they need a supply of oxygen. In big complex animals this is possible because of specialized breathing organs—gills or lungs—and also specialized transport systems for getting the oxygen around the body. But the very first animals (and indeed also today's simplest animals) had (or have) no such specialized organs or systems. So how did they cope?

The answer lies in adopting a thin body form, one in which no cell is too far from the water or air in which the animal lives. Nobody can say exactly what the first animal looked like, but some simple creatures still around today give us important clues. They take the form of small hollow sacs of cells. They may be spherical or flattish—like a new, fully inflated beach ball (in miniature) or an old, punctured one. Their skin is just a monolayer of cells, so all of these are in direct contact with the external environment, though there may also be a few internal cells that are able to acquire oxygen by diffusion through the thin outer skin.

These primitive creatures, of which there are several kinds, have just one or two cell types, in contrast to our own two or three hundred, and a total of anything from tens to thousands of cells (depending on which group you're considering), in contrast to our trillions. But they work. As functioning organic machines, they go about their simple business and survive. Some have survived so well using their simple body form to conduct the basic tasks of life that they have remained almost unchanged in perhaps a billion years. Others not only survived but also, through further duplication and divergence, led to more complex creatures.

What exactly are the basic tasks of life? We've already looked at breathing, taking in oxygen, and, implicit until now, getting rid of that waste product of breathing, carbon dioxide. What is true of breathing is also true of feeding—something useful goes in, something not so useful comes out. Again, these processes are facilitated by a thin body form. Movement in tiny primitive animals can be passive, so no muscles are needed, and thinking is nonexistent. But there is one other basic task of life that even the simplest animals must be capable of: reproduction.

It's all too easy to think of animals in only three dimensions. I usually think of my brother as he looks now. I rarely picture him as his much younger, childish self. More rarely still do I think of him as the old man he will one day probably become. And I find it very hard indeed to picture him as the hollow ball of cells that he was in our mother's womb a long time ago—a ball that was curiously reminiscent of the simple animal we have been discussing for this last while.

Yet all animals are four-dimensional things. As I, and others before me, have said, animals do not *have* life cycles; rather, they *are* life cycles. And now we need to think for a moment about the life cycles of those earliest mini-beach-ball animals.

Notice that I have, without really announcing the fact, added development to the picture. We began our time-extended view of animals with reproduction, but we have ended up with life cycles. A life cycle includes, at minimum, both reproduction and development. It also includes aging and death; but let's leave aside those depressing subjects and focus on the creative side of life cycles.

So our beach-ball animal must be able to do two more things: detach a cell from somewhere around its periphery, letting it float gently away from its parent, and program that cell so that it will proliferate daughter cells that stick together as they grow into another hollow-sac adult. Detachment means temporarily switching off the gene that makes the sticky protein; programming subsequent development means switching that gene back on again. Evolution involves changes in the ways in which genes are switched on and off, as well as changes in the nature and number of genes.

All this is hard to picture because the earliest animals were very small—probably invisible to the naked eye. And their closest counterparts among today's fauna are also small. But some of today's simple animals have grown much bigger, and so are easier to picture, despite remaining at a very low level on the ladder of complexity. The best known of these are sponges and jellyfish.

The problem with sponges is not so much picturing them as remembering that they are indeed animals. Those of us who are reasonably long in the tooth will have used the softish skeletons of real sponges to wash ourselves with in the bath. Younger folk will also be familiar with ablutionary "sponges," even though these are mere plastic fakes. Anyone lucky enough to have snorkeled through warm, shallow coastal seas will have seen live sponges growing on the seabed. They don't do many of the things that more complex animals do—they don't swim, walk, or growl. But they are quietly feeding and breathing, and from time to time reproducing.

Adult sponges can be quite large, up to a meter or more in their longest dimension, so they consist of a large number of cells. Perhaps the largest sponges pass the million mark. Yet the number of cell types is small. Estimates vary, and of course the number of cell types in any creature depends on how finely you subdivide them. At any rate, there are at least three quite different types of cells in a sponge, and one of these—the choanocyte (or collar cell)—has some similarities with a flagellate protozoan, which is why many biologists reckon that this particular protozoan group may be the one that contained the ancestor of animals.

Moving up from the cellular to the organismic level, we can see that sponges have the following features—or lack of features: They have no tissues or organs as such, so, for example, there are no muscles and no brain (or indeed any nerve cells at all). Their body is asymmetrical, so there are no body axes, such as anterior-posterior. Like all sessile animals, they reproduce via a larval form that enables dispersal by floating away from its parent and settling elsewhere on the seabed, where it will grow into an adult.

Sponges function simply by wiggling little hairlike projections (the flagella that stick out of the collar cells) and thus creating water currents: water flows in through small pores in their bodies and out through larger ones. Both feeding and breathing occur by taking things from, and giving back other things to, these currents. That's about it. These are indeed the simplest of all large animals.

And so to jellyfish. We humans have a curious tendency to label a lot of sea creatures, and indeed some freshwater ones, too, as something-fish, even though many of them are not fish at all. The impostor groups include jellyfish, starfish, crayfish, and the catchall gastronomic shellfish. In contrast, actual fish groups include monkfish, anglerfish, and dogfish. It's worth keeping in mind that "fish" can mean just about anything as an attachment to some other word. Jellyfish, our subject here, are about as distantly related to true fish as they possibly could be: they are, if you like, at the other end of the animal kingdom.

Jellyfish are impressive creatures. They can reach considerable sizes: some are larger than humans. All have stinging cells that they use to attack prey. Some of the biggest ones, with the most effective stinging cells, can be dangerous to humans, including the wonderfully named Portuguese man-of-war, which I was taught to watch out for as a child, when swimming off beaches where they made an occasional appearance.

Large and impressive as they can be, jellyfish are still at the simple end of the spectrum of animal complexity, yet they have made several advances over their sponge cousins.

First, they have more cell types. Again, putting a precise figure on it is difficult, but there are several kinds of cells that jellyfish have in common with more advanced animals and that sponges lack. Perhaps the most important of these are nerve cells. Although jellyfish have no brains, they do have what is called a nerve net: a network of interconnecting nerve cells that extends through most of their body. In some cases, these networks are very diffuse; in others, nerve cells are concentrated in certain places—either around the mouth or in the form of rings around the animal's outer margin. The nerves serve a variety of functions, including sending signals to the muscle cells that power the primitive swimming movements that jellyfish can make.

Second, they have moved from an irregular body form to a regular one. Their particular kind of regularity is often referred to as radial symmetry, because if you were to slice a jellyfish in two from top to bottom, making sure your cut is a true diameter—that is, it goes through the central point—the two halves of the jellyfish would be mirror images of each other. And this is true regardless of the angle of your cut, in the sense of north-south versus east-west, looking down at the jellyfish from above and thinking of it as a compass.

Third, jellyfish are often referred to as two-layered animals. At first sight, this does not seem to make much sense. But it refers not so much to large-scale features that are visible in the adult as to small-scale embryonic features that are invisible to the naked eye.

Most animals, ourselves included, are described as three-layered. The embryonic layers are called the ectoderm, mesoderm, and endoderm—which translate as outer skin, middle skin, and inner skin, though I suppose "tissue" would be a better word than "skin" here, as skin is usually thought of as an outermost tissue, almost by definition. As a human embryo develops, the different layers give rise to different things. For example, the skin is indeed formed from the ectoderm, the muscles from the mesoderm, and the gut from the endoderm. I didn't mention these layers in relation to sponges, because they don't have any. Jellyfish have ectoderm and endoderm but lack mesoderm, hence their description as two-layered—though some recent work challenges the accuracy of this description.

So jellyfish are intermediate between sponges and the rest of the animal kingdom in several aspects of the complexity of their bodies, measured, as ever, by the number of different types of component parts. This is true both at the level of the cell and at the higher level of structures made of cells, such as nerve nets and tissue layers.

I titled this chapter "From Simple to Complex." Whether a jellyfish is complex or not is, of course, a moot point. It can be argued that describing any one animal as simple or complex is meaningless. These terms only make sense in a comparative context. A biologist who has just completed a major research project on jellyfish may think of them as amazingly complex creatures. After all, there is a wealth of structural detail (most of which I have passed over here) in even the smallest and simplest jellyfish, and the number of types of overall signal patterns that can flow through even a very basic nerve net is huge. On the other hand, a vertebrate zoologist who has spent a lifetime studying the most complex of all animals may well think of jellyfish as incredibly simple. Both viewpoints are fine—it all depends, as they say, on "where you're coming from."

There is one final point I need to make before moving further in the direction of increased complexity as we continue to ascend life's ladders in the next chapter. This point relates to some close relatives of the jellyfish that I haven't yet mentioned.

Zoologists divide the animal kingdom into about thirty-five large groups. The sponges are one of these. Their official name is Porifera, which means pore-bearing (or hole-bearing, if you prefer), this being a reference to that ubiquitous feature of the group, the presence of lots of holes over the surface of the body through which water flows in and out. Here the common name, sponge, and the scientific name, Porifera, have a one-to-one correspondence. To put it another way: all sponges are poriferans and all poriferans are sponges.

The same neat one-to-one correspondence between common

and official names does not apply in the jellyfish world. In this case, the large group to which jellyfish belong, equivalent to the sponges' Porifera, is Cnidaria (with a silent *C*; called after the stinging cells, or cnidae). But this group also includes several close relatives of the jellyfish, such as sea anemones and corals.

Much of what I have said about jellyfish applies to these creatures also—for example, few cell types and (probably) only two embryonic tissue layers. So our complexity story is unaffected by the existence of these superficially different but fundamentally similar relatives.

But here is an interesting fact about complexity of life cycles. Sea anemones, which many of us recognize from visits to coastal rock pools when we were children, are rather plantlike animals, in that they do not move—they are attached to the rock. Many other cnidarians have this same sessile existence and look a bit like sea anemones—for example, freshwater hydras, which are generally smaller and less colorful than anemones. This general body form—attached, sessile, mouth and tentacles pointing upward—is called a polyp. (The same name is given to some cancerous growths in humans, as they have a roughly similar form.) The jellyfish is in many ways opposite to this in design: mobile, unattached, mouth and tentacles pointing downward. This opposite body form is called a medusa.

Now, here's the crazy thing that is unknown to most nonbiologists. In many species, the life cycle consists of an alternation of polyps and medusas. That is, a thing that looks like a sea anemone gives birth to a thing that looks like a jellyfish, which completes the cycle by giving birth to a thing that looks like an anemone again. Perhaps this is no more remarkable than other complex life cycles, like caterpillar-butterfly or tadpole-frog, but it is certainly less widely known.

CHAPTER FIFTEEN

FROM COMPLEX TO EVEN MORE COMPLEX

We find it easy to think of ourselves as the pinnacle of evolutionary complexity. In terms of brain complexity, and resultant behavioral complexity and intelligence (see Chapter 17), this viewpoint may be a reasonable one. But in terms of complexity of overall body form, evolution has not one pinnacle but many. All our fellow vertebrates, both terrestrial and aquatic, have body forms immensely more complex than those of the simple sponges we looked at in the previous chapter. And many invertebrate animals, notably among the arthropods and mollusks, also have impressively complex bodies.

It is often said that human civilization began in Mesopotamia. This name means "between two rivers"—specifically the Tigris and the Euphrates. The area concerned is, ironically given recent world events, largely within the modern country of Iraq. One of the main cities in Mesopotamia—and for a while its capital—was Ur. So, in a sense, Mesopotamia in general, and Ur in particular, were the start of things. Given this, we might consider using "Ur-" as a prefix meaning "the original." Strangely, the English prefix "ur-" (note the lowercase *u*) is used in this way but for an entirely different reason: it has been imported from German, where, for example, the original version of a text is referred to as the *Urtext*.

Our focus here is on animals, not texts. So, enter the animals that have been called the urbilaterians. Exactly who or what they were is the question.

We have already seen that early animal evolution led from an irregular body form lacking any symmetry (sponges) to a body form with radial symmetry (jellyfish and their close cousins). The next step was to go from radial to bilateral, or two-sided symmetry. This kind of body has a head end, a tail end, a left side, and a right side. If you slice such an animal in two, the left and right halves will be mirror images of each other. But unlike with the radial jellyfish, slicing a bilateral animal requires control over the angle of cut. The right and left sides of a rat, for example, are mirror images, but its front and back are not. Perhaps I should add that I am thinking of this slicing as a mental, not a physical, exercise . . .

Anyhow, the important thing is that nowadays almost the entire animal kingdom is based on a bilaterally symmetrical design. Sponges and jellyfish can be thought of as early evolutionary side branches that split away from the rest of the animals before bilaterality was "invented." The creature that came up with this invention was the urbilaterian.

What do we know about this animal? The short answer is, "Very little." We have no fossils of it to examine. We can only make inferences about it from indirect sources of evidence.

One such inference is that it lived a very long time ago. Fossils from the Cambrian explosion more than 500 million years ago include many with bilaterally symmetrical body forms. And indeed some of these are quite elaborate and look like they have already had a substantial period of bilateral history. There is currently considerable debate about how much earlier the urbilaterian was, compared with its Cambrian offspring, but that need not concern us here, as we are focusing on the nature of evolution's upward steps in complexity. This focus requires getting the time *sequence* right, but it does not require estimates of absolute time to be right, too.

There is also much debate over whether the urbilaterian was a segmented animal, and whether it had any kinds of limbs. My view is that it probably had neither of these features. But in any event, I want to concentrate on one feature that it most certainly did have—the one that gives it its name—bilateral symmetry.

You might well ask why bilateral symmetry is more complex than radial. Let's make a particular comparison: between a jellyfish and a flatworm. If you haven't seen a flatworm, here's what it looks like: take an earthworm, shrink it down to a length of only half an inch, gently stand on it to flatten it (mentally again, of course), imagine that it has no visible segments, and change its color from pink to brown. Oh, and one final thing: think of its head end as slightly wider than its tail end, so that it tapers from front to back. (The reason I chose a flatworm here is that some particular present-day flatworms look like good candidates for being the closest living relatives of the urbilaterian.)

Now, forgetting about internal detail like cells, and just focusing on the overall body shapes, is either more complex than the other? Does a flatworm have more different types of parts than a jellyfish? Well, not really. Yes, it has left and right sides, which a jellyfish does not. But that's about it. Although the flatworm has head and tail ends while the jellyfish does not, the jellyfish has a topside and an underside (the latter having the mouth), which pretty much amount to the same thing. And the jellyfish has tentacles, while the flatworm does not.

So now we are getting close to the main message of this story. The evolutionary shift from radial to bilateral designs was not really a step up the ladder of complexity. But it was a shift that *led* to many such later steps.

This bold statement requires some explanation. Let's start poking at the issue from an ecological angle. Jellyfish swim in the open seas. Their sea-anemone cousins remain rooted to the rock. Neither can move on a solid surface, whether by gliding or creeping. But that is exactly what flatworms do. And it is probably what the urbilaterians did hundreds of millions of years ago. Think of them as tiny marine slugs oozing their way across boulders in the coastal lagoons of ancient tropical seas.

Simple creatures but with (unknown to them) much potential for later complexity. And the reason: forward movement. For a jellyfish there are no such things as forward, backward, and sideways. With a radial body form, they are all the same. So, apart from having a top and an underside, no part of the animal needs to be any different from any other. And that's fine. After all, jellyfish have been remarkably successful. As a group, they have existed for more than 500 million years. There are thousands of different jellyfish species in today's oceans—some of them occurring in great abundance. Their deadly stinging cells enable them to kill and consume creatures, such as fish, that are much more complex than they are.

But, again unknown to them, they were a sort of evolutionary cul-de-sac. Recall their brainlessness. Their nerve networks are very diffuse affairs. Their behavior patterns are very limited.

But now think again of that primitive urbilaterian with its "head" at the front, gliding along a rock. Suddenly there are a whole lot of reasons why the front end should be different from the back. It is more important to be able to detect the features of the area you are moving into than the one you are leaving behind. So it is to be expected that natural selection will favor variants that have more of their nerve net up front. As part of this concentration of perceptive powers at the front of the animal, in the region that will one day merit the term "head," there may be rudimentary forward-pointing outgrowths densely populated with sensory nerves—primitive antennae.

Like having a concentration of nerve cells at the front, having a mouth at the front also makes sense. Forward movement facilitates the engulfing of things that lie in your path, not things that are by your side or behind you. And this business of engulfing requires some control—another reason for having a concentration of nerve cells at the front. Many of today's simpler bilaterian creatures, such as slugs and snails, have an arrangement of mouth and nerve cells such that the nerves form a ring around the mouth or the esophagus, and at various points around this ring are expanded groupings of nerve cells called ganglia—or mini-brains, if you like. We already saw, in Chapter 12, that this simple kind of arrangement in primitive mollusks formed the basis for expansion into the impressively complex brain of the octopus.

So, much of the evolutionary potential lurking in those simple urbilaterians was connected with their forward movement. But some of the subsequent steps up life's ladders were taken for quite different reasons. I'd like to concentrate on one of these: increased size.

It seems that most animal groups launch themselves on the evolutionary scene in miniature. The first animals were tiny. The first jellyfish were hardly Portuguese men-of-war. The urbilaterian flatworms were probably smaller than a paper clip, albeit this is guesswork because, as I said earlier, we have no fossil remains of them to study. And in later evolution, the first mammal was probably closer to the size of a shrew than to that of an elephant.

In a group like mammals, where the creatures concerned are already very complex at the outset, increased size need not cause any pressure for a redesigned body. All the existing parts can simply be scaled up. But in a group like urbilaterians, which were much simpler, increased size brings with it a host of problems. The solutions to these problems generate complexity.

Perhaps the best two things to focus on in relation to this point are systems for breathing (acquiring oxygen) and transport (getting it around the body). The urbilaterians probably had neither of these kinds of systems. They didn't need them, because they were small. No part of the animal was too far from the external environment, so everything could happen by passive diffusion of gases. But the bigger you get, the less well this works; and past a certain size, it doesn't work well enough to keep you alive. As a creature approaches this critical size, natural selection in favor of any variants that have features helpful to breathing and transport will be strong.

The first such features were probably quite unimpressive compared, for example, with the human lung or heart. But that, of course, is irrelevant. Natural selection was, in the urbilaterians as elsewhere, dealing entirely with the present, oblivious to the future. And unimpressiveness is in some ways a blessing in disguise, because the simpler the feature, the easier it will be to create it through the accidental process of mutations and duplications of genes, and the subsequent reprogramming of the developmental process through which the genes make the animal.

Let's make a brief digression from breathing to typing. I am typing this text with, as you might imagine, my fingers. (I have nothing but awe and respect for that famous disabled Irishman, Christy Brown, who typed a book manuscript with the toes of his left foot.) The reason we humans have unconnected fingers and toes, in contrast to, for example, the very much connected digits of a duck's webbed foot, is that during embryonic development the cells in the interdigital regions die.

A rudimentary gill slit could be made in a similar way: the death of rows of cells, this time in the head region, leaving slits. Perhaps this was indeed the way gills began, perhaps not. But the basic point remains the same: simple beginnings can easily be produced by accidental changes in the developmental process, and they can later be refined in many directions. So we go from no structure to rudimentary structure to elaborated structure. From simplicity to complexity.

Gills help an aquatic animal to acquire more oxygen from its environment than it can by diffusion alone. But they don't solve the problem of getting it around the body. That job, which in a mammal is accomplished by our well-developed cardiovascular system, is in most simpler animals carried out by a kind of primitive precursor of this system. At its simplest, this precursor is just a body cavity within which mobile blood cells can float around in a fluid, nourishing whatever parts they come into contact with. A cavity is something that can easily arise when the course of development is kicked into a different route by the mutation of a gene that helps to control it. And again, once something has begun, selection will favor refinements of it that work more effectively. So, with increased size, it helps to have blood being powered around the body rather than merely sloshing around it passively.

The step from blood cavity to blood vessels requires the construction of tubes. This step seems like a big one, and big steps are often seen as problematic for an evolutionary process driven by natural selection. But on inspection this big step is not quite as big as it seems. If a primitive bilaterian animal already has a mouth leading to an esophagus, then its body "knows" how to make a tube. It has the necessary genetic machinery to achieve this feat. So all it has to do is redeploy this machinery in a different context.

We're getting into a very exciting area here—a process that goes by the name of co-option. Genes that are already in the animal doing one job can be co-opted to do another, related job. The importance of this process has only really become apparent in the last few years, and there is much yet to be learned about it. But perhaps it is as important, in its own way, as duplication and divergence are, as a contributor to the ascent of life's ladder from simple body forms to complex ones.

Let's recap. Here are several upward steps in complexity, which took many millions of years to achieve, condensed into a single paragraph: from unicell to multicell; from irregular body form to symmetrical one; from radial to bilateral; from no head to proto-head; from no organs of breathing or circulation to the beginnings of such organs. Then: the elaboration of all of these, resulting in gills, lungs, hearts, and brains. All of this is based on gene mutation, developmental reprogramming, and natural selection; and often on specific versions of these such as gene duplication, divergence of replicated parts, and co-option. Magnificent and diverse results from simple, natural processes. As Charles Darwin said in his *On the Origin of Species*, "There is grandeur in this view of life."

In the previous chapter, we saw that the process of going from simple to (a bit) complex was based on the ability of cells to stick together. In the present one, we have seen that the process of going from a bit complex to more complex was based on a bilateral design, and that further complexity still was ushered in by the challenges posed by large body size. Let's now look a little more closely at these two key features—bilateral symmetry and body size. Among other things, this closer look will help us avoid the ever-present trap of taking away from the story too simple a message.

I keep reemphasizing the fact that as an accidental process, evolution is inevitably messy. We see patterns, such as increased complexity, but they are untidy patterns that are in a sense statistically rather than absolutely true. That is why I have been careful throughout to draw attention to the fact that it is just the *average* complexity of creatures that has increased over evolutionary time. Biology is not like physics. There is no law that says all organisms must climb life's ladders. Many have not climbed them at all, and many have slid down them to simpler forms than those that characterized their ancestors. These points about messiness and reversibility apply to bilaterality and body size just as they do to other features.

Although becoming bilaterally symmetrical was a great leap for animalkind, many complex animals that have acquired their complexities partly in consequence of their ancestors being bilaterians have either muddied their bilateral symmetry or dispensed with it altogether. As mammals, we are very nearly symmetrical from an outside view. This symmetry is imperfect, of course: most people on close inspection can be seen to have a bent nose, hands that don't exactly match in size, or one ear sticking out a little more than the other. But these departures from perfect bilateral symmetry are as nothing compared with the much bigger departures that would be seen from an inside view—one rarely afforded us, except indirectly through pictures. We do not have a centrally located heart or two separate hearts, one on the left and one on the right. Rather, as is well known, we have a left-side heart. (Occasionally, as part of a condition known to the medical world as situs inversus, it is on the right.) And various parts of our digestive system, too, depart significantly from bilateral symmetry. In other kinds of animals, other departures from bilaterality can be seen, both outside and inside. Some crabs have one pincer much bigger than the other. Snails' shells spiral to the right or the left. They do not have mirror-image left-hand and right-hand shells—though their close relatives the bivalves do, with varying degrees of imperfection.

Ironically, many of these departures from perfect bilateral symmetry are themselves steps up the ladder of complexity, because if left-side and right-side structures are different, then the creature concerned has more different kinds of component parts than it would have had if it had remained perfectly bilaterally symmetrical. But some animals have dispensed with bilaterality completely, and have tended to revert to lower levels of life's ladder than those animals that retained a bilaterally symmetrical body plan.

The most striking example is that strange group of creatures called echinoderms, or "spiny skins." Included here are starfish, sea urchins, and some lesser-known relatives, including the wonderfully named sea cucumbers. Echinoderms are usually said to have a special form of radial symmetry—pentaradial. If you go down to the coast and pick up one of the dead starfish that can often be found lying among the rocks, you will probably find that it has five arms. There are many species of starfish, and most of these do indeed have five. Those that depart from this number often have some multiple of five, such as ten or fifteen, though there are a few of which this is not true. Since starfish do not have a head or a tail end, and hence have no forward direction of movement, they have a diffuse nerve network rather than the concentrated kind that includes a brain. But we know from two kinds of evidence (fossils and molecules) that starfish are secondarily radial. That is, unlike jellyfish, they had bilaterally symmetrical ancestors.

And so to body size. Many of the great groups of animals exhibit incredible variation in size among their constituent species. Sometimes, as in mammals, this variation covers several orders of magnitude. And while most groups begin with smallish body forms and evolve upward in this respect, this trend, like that in complexity, is statistical rather than absolute. Many lineages within a group that is evolving upward in average body size buck the trend and go in the opposite direction.

Horses are a classic example of evolutionary increase in body size. The first ones were closer in size to dogs. But hummingbirds are much smaller than the ur-bird (to coin a word) that arose from a dinosaurian ancestor. Getting smaller often does not mean dropping down the scale of complexity. Hummingbirds and blackbirds don't really seem any different in complexity terms.

So much for caveats and complications: back to the main theme. Animals, and plants, too, for that matter, have increased in average complexity over the eons. Some of this increase has been a slow and quiet affair. But some steps can be seen as of crucial importance. One of these, a bilateral body form, I have focused on especially in the present chapter.

It is bilaterality that bestows upon a creature a head end and a tail end. As we have seen, the enhanced development of the head end has been a major contributor to the increased complexity of many animal groups. Where the head goes, the rest of the body follows, in more ways than one—albeit we upright humans represent a departure from this pattern in terms of our bipedal manner of getting around.

Given the importance of the head, we will now focus on it entirely for the next two chapters. We will look in turn at its inception and its evolutionary apex, and, in the latter case, at the incredible behavioral consequences of crossing some invisible threshold in the size and complexity of the brain.

CHAPTER SIXTEEN

ACQUIRING YOUR HEAD

Most of the animals we see around us have heads: our fellow humans, dogs, cows, frogs, butterflies, even snails. But we all arose, in two ways, from headless beginnings. We have acquired our heads twice—once in the course of evolution, and once in our own embryological development. Here we look at these dual acquisitions.

LUCKILY, the expression "losing your head" is not normally used in a literal manner. Rather, it refers to someone losing his or her temper and behaving on the basis of emotion—usually anger—instead of reason. But losing your head in a more real sense is, of course, entirely possible. The ancient art of guillotining springs to mind.

Those humans unlucky enough to have literally lost their heads, such as members of the French royal family after the Revolution, were killed instantly. A pathologist called in shortly after the event would have had little problem in certifying them as dead. Contrast this with the situation lower down the animal kingdom. Earthworms are famous for their ability to regenerate. If your spade accidentally cuts one in half as you drive it vertically into the soil, perhaps in an attempt to neaten up the edge of your lawn, the remaining head end will grow a new tail, though the tail end will not grow a new head. However, in flatworms (very distant relatives of earthworms), regeneration of heads occurs with ease.

Some animals can even function reasonably well without a head. I was once collecting centipedes from under stones near a stretch of coastline in northeast England. I took these back to the lab for closer inspection. They seemed healthy and went about the business of investigating their new environment in a normal way. On microscopic examination, some could be seen to be missing legs, or bits of legs, but, with the abundance of legs that centipedes have, this was not a problem. In fact, they have a clever trick of covering the stump in a black substance that looks suspiciously like tar (but isn't), and which perhaps acts as a seal that prevents both leakage and infection.

To my surprise, I found that one specimen had lost its head and had used that same black-tar technique to seal the wound. Strangely, it was moving around in a normal fashion just like all its friends who had retained their heads: it walked in the way that centipedes do, usually in a forward direction, but was also able to retreat backward from threatening stimuli (such as my giving it a tap on its anterior end with a pair of tweezers). Not only did it do so for a considerable period in the lab (hours), but it had probably done so for an even longer one (days) in the wild before I found it, because the wound looked old, inasmuch as you can tell these things.

Now, this centipede no doubt eventually died of hunger, because, lacking a head, it had no mouth. But until such time it functioned fine. The lack of a brain was not a problem, because centipedes have very small brains; and they also have bundles of nerve cells in each segment. These bundles can do a lot on their own without the assistance of the slightly bigger bundles—that is, the brain—at the head end of the creature.

The contrast between this centipede's continuing life after losing its head and the instant death of humans who have been guillotined hinges on the degree of cephalization in these two very different creatures. We humans have become much more cephalized—that is, we have much more of our bodily control centered in our heads than is the case with the lowly centipede.

I will refer to the process of cephalization as "acquiring your head." We each have acquired our heads in two different but related ways: evolutionary and embryonic. I'll take them in that order.

In the previous chapter, we saw that early animals went from no head (such as jellyfish) to proto-head (the so-called urbilaterian, probably a kind of simple flatworm). Now we must pick up this story where we left off and consider the transition from proto-head to proper head.

Of course, this presupposes that we know what a proper head is, and this issue is worthy of a little attention at the outset. That early, first-of-all flatworms, which crept across the surface of rocks submerged in ancient seas, probably had just a slight concentration of nerve cells in its head region, coupled with some very simple sensory apparatus—maybe light-sensitive pigment spots and some kind of chemical-sensing cells.

Contrast that with creatures possessing more advanced heads—not just humans and their vertebrate cousins but also insects, crustaceans, mollusks, and many other animal groups. We see a tremendous variety of impressive head structures in these creatures, both those visible from the outside and those that would only be visible from within. Together, these include brains, eyes (simple, complex, and compound as in insects), antennae, tentacles, ears, noses, mouthparts (including jaws), and devices for echolocation (as in bats).

So creatures that possess at least some of these many features can be said to have proper heads. Of course, there is no clear line separating proto-head from proper head, and, as with most evolutionary processes, there is a gradual transition from one form to another.

But we again need to remind ourselves here that evolution is not an escalator on which everyone travels upward at the same speed. Indeed, not only do some creatures travel upward slowly, others fail to get moving at all, and still others travel in a downward direction. Those spiny skins, the starfish and their kin, which dispensed with bilaterality and went back to a form of radial symmetry of their bodies, have, effectively, lost their heads. People working on starfish tend to label them as having not a head-to-tail body axis but a mouth-to-without-mouth axis.

If we leave the stick-in-the-muds and the sliders-downward aside, and concentrate on those many creatures that have indeed climbed up life's ladder from proto-head to proper head, there is another crucial point that needs to be focused on before we go any further. This concerns the fact that, as noted earlier, life has not just one ladder but many.

Perhaps the best way of approaching this is to combine the usual tree metaphor for evolution with the ladders I am using here. Think of all those bilaterian animals that have evolved upward in head form from the starting point of their ancient urbilaterian common ancestor. Their pattern of relationship takes the form of a tree (or a bush, where there is explosive evolutionary radiation). Their ascent in complexity—in this case in terms of the head—takes the form of a ladder. Put the two mental images together and you get a tree whose branches are ladders.

Now we begin to get a grip on what is really going on. Any two animals, whether humans and horses or lice and lobsters, go up the same ladder for a while, and then they diverge to ascend different ones. The degree to which their heads are a shared evolutionary invention, as opposed to an independent experiment in cephalization, is determined by the extent to which they travel upward together before parting company. In the two pairs of examples I just gave, the common climbing was considerable. But if we rearrange the four animals concerned, so that the pairs become humans and lice, horses and lobsters, then the degree of common ascent before the parting of the ways was much less.

Suppose that two lineages diverge at a point where the last common ancestor has a brain but no eyes. Later in evolutionary time, if both have acquired eyes, they have done so independently—a phenomenon referred to as convergent evolution. But the brains that reside in the heads of creatures of the two lineages are not independent, at least in their basics, which were inherited from that last common ancestor—this is referred to as homology. Subsequent elaborations of the brains (after the lineage split) were, again, independent.

The two things—homology and convergence—are not always easy to separate. It is clear from a comparison of human and insect eyes that most or all of their evolution has been independent. But a comparison between human and octopus eyes is less easy to interpret. There are impressive similarities between the two, and it requires a very close look at their structures to reveal the telltale signs of convergent evolution.

It's time to focus in, but not so much on eyes as on that most fundamental head structure, the brain. And, given the tree-of-ladders picture, we need to focus on one particular ladder in order to remain sane. I'm going to succumb to the lure of a "vertebratocentric" approach on this occasion, and focus in particular on our own large group of animals. Had I chosen instead to focus on mollusks or arthropods, the story would be the same in its essentials but different in its details.

The most primitive vertebrates around today are the jawless fish—for example, the lamprey. They represent the opposite end of the cephalization ladder to humans, though we must, as ever, be careful here not to confuse cousins with grandparents. Present-day lampreys fall into the former category. We often think of them as grandparent look-alikes, in that they have changed much less, since the lineages split, than we have.

But our vertebrate story needs to begin a little earlier than the first appearance, in evolution, of jawless fish. It always makes sense to have a look at a close relative of the whole group you are focusing on (an out-group, as it is often called), as well as that group itself. In the case of the vertebrates, one of our closest non-vertebrate relatives is a small, slim-line sea creature called the lancelet (after its shape) or, alternatively, the amphioxus (also referring to its shape, as this word translates to "double-pointed").

We are all familiar with our own backbones. We can feel their constituent vertebrae if we run a hand down our back. We know, from museum skeletons of bears or whales or birds, that all our fellow vertebrates also have backbones. Indeed, this is where the name of our group—the vertebrates—comes from. The little lancelet has no backbone, but one of the pointers to its being a close relative of the vertebrates is that it does have, running down the center of its back, a stiff rod called the notochord, which looks for all the world like a backbone-in-waiting. In fact, a human embryo has a notochord for a while. It acts as a signaling center that tells the surrounding tissues to make a variety of things, including the backbone that eventually replaces it.

The group to which the lancelet belongs is called the Acrania. The cranium in a human (or other vertebrate) is another name for the main part of the skull—the braincase. Lancelets don't need a braincase, because they don't have much of a brain. But they do have a dorsal nerve cord, apparently a forerunner of our own spinal cord (again with the cousin-versus-grandparent proviso), that is slightly swollen at the anterior end—a proto-brain, as we might want to think of it.

The existence of this dorsal nerve cord is especially interesting because most other bilaterian animals, whether insects, snails, worms, or whatever, have a ventral nerve cord. Indeed, it appears that in the evolutionary divergence of the bilaterians, one of two separating lineages in the very distant past became inverted in relation to the other. The French zoologist Étienne Geoffroy Saint-Hilaire came up with this theory in 1822 from his purely anatomical studies, and many of his contemporaries dismissed it. But in the last couple of decades, new genetic support for his theory has been forthcoming, and it is in vogue once again.

The lancelet's anterior sense organs are just as rudimentary as its brain. It has a tiny light-sensitive pigment spot at the anterior end of its nerve cord. This spot serves as a very primitive eye. The lancelet has no ears and almost certainly cannot hear. It has a series of anterior projections that may have some sensory capability, and it has a series of gill slits, still at the anterior end but a bit farther back. And that's about it.

Turning now to the vertebrates themselves, we see a vast range of degrees of development of the head in general and the brain in particular, ranging from the jawless fish to their jawed counterparts to the land vertebrates from amphibians upward. In general, reptiles are brainier than amphibians, birds more so than reptiles, and mammals the brainiest of all. But as ever such generalizations are imperfect. The brightest birds can learn more complicated tasks than the dumbest mammals. Anyhow, we'll pick up this story again, in rather different ways, in the next couple of chapters.

Before moving from the evolutionary side of acquiring a head to the embryological side, I need to make one final point. Although the head is special because, in its human form, it is the very thing that enables us to think about issues like our own origins, its elaboration from simple beginnings requires no special explanations, over and above those given for the evolution of complexity in general in previous chapters. So in general terms the explanation lies in the accidental supply of variation through copying errors in the genetic material, the reprogramming of development as a result of these accidents, thus taking new variants from the purely genetic level to that of the whole organism, and the sifting of the variants by Darwinian natural selection. The specific version of this combined process that is most important is, in head elaboration as elsewhere, duplication and divergence, with the latter—that is, the divergence—often taking the form of co-option; that is, a redundant copy of a duplicated gene is co-opted for a new role in the control of development.

And so to development itself. Although humans could be said to be the most cephalized creatures of all, when we begin life we have no head—not even a hint of one. Recall the connection between the short-term changes in embryos and the long-term changes in evolution. We saw, in Chapter 9, that, albeit in a very imperfect way, the embryogenesis of a higher animal often repeats in condensed form (or recapitulates) those evolutionary changes by which it arose from a simpler ancestor. This recapitulation applies to acquiring a head, just as to acquiring other features.

Like most animals, we begin life as a single cell—a fertilized egg. No sign of a brain there, of course. After the first rounds of cell proliferation, we are little balls of cells that are all rather similar to each other. So, still no sign of a brain, or even a rudimentary head. Later, when the embryo grows further and undergoes some rather radical shape changes to become more elongate in one axis than in the other two, the discerning microscopic observer will be able to point to one end of this main axis and exclaim "Head!" with excitement. And later still, as the dorsal midline of the embryo goes through a burst of developmental activity, more pronounced at the head end than the tail, a sort of primitive central nervous system, complete with proto-brain, becomes apparent.

All these things take place in the first couple of weeks post-fertilization when the embryo is still very small. It may, by the end of this time, consist of thousands of cells, but it is only just about big enough to be visible.

There are still many weeks of embryogenesis to go, plus many years of post-embryonic growth. During this vast span of time for a developing creature (but only a split second for evolution), much further elaboration of the brain takes place, at the end of which we have moved from the basic perceptions of the newborn to an ability to handle abstract concepts and theories, whether the weather forecast or trigonometry.

Like the evolutionary acquisition of a head, its developmental counterpart is based on natural mechanistic processes rather then magic. Evolutionary processes like natural selection and co-option are replaced by developmental ones like the switching on and off of genes in particular parts of the embryo at particular times. These are the "dances with genes" that result in the building of a "castle of cells," to bring back those vivid mental images from earlier chapters.

I suppose that my telling of the story of cephalization, in this chapter and the next, is biased in a particular way. It emphasizes the beginning of the process, and also its end, or culmination, inasmuch as there is such a thing.

This bias is deliberate rather than accidental. What happens in between these extremes of beginning and end is essentially "more of the same." Whether we are thinking about the evolutionary beginnings of our brain in the slightly swollen anterior end of the primitive nerve cord of a lancelet or about the developmental beginnings in the early embryo, the processes of elaboration that occur to move us up the first rung of the cephalization ladder are merely repeated, with some variation, to take us up subsequent rungs. One of the key things, in both development and evolution, is increase in cell number. This is true regardless of which group of animals we are considering. If we abandon our vertebrate focus for a moment and think of the brain of an octopus, increase in cell number was crucial to evolution in making the first octopuses from snail-like ancestors; it was also crucial to development in making intelligent adult octopuses from their unthinking embryos.

All this sounds like a very gradual, cumulative process. And so, most of the time, it is. If you like, the steps of life's ladders seem to be many and very close together, so that any particular step upward is barely discernible. But little about evolution is neat and tidy, as I have stressed on many occasions already. Life's ladders differ from real ladders not only in having their rungs closer together but also in having the space between rungs variable rather than constant.

The more widely spaced rungs of evolution are interpreted very differently by different groups of people. To the creationists they represent an opportunity for distortion into an apparent impossibility that testifies to the falsehood of evolution in general (see Chapter 20). To the ultra-gradualists of evolutionary science, like Richard Dawkins of *The Selfish Gene* fame, they are an annoyance because they seem to be (but are not) threats to a Darwinian worldview. But to those of us fascinated by the new "evo-devo" approach to evolution, they are things of particular interest. They seem to indicate points in evolution where something really interesting happens.

Sometimes, the really interesting thing has to do with evolution of the developmental system itself. But sometimes, the interest lies not so much in that, but rather in consequences that flow from it. And that is the case in relation to the human brain. If we trace a particular vertebrate lineage all the way from proto-fish to humans, the size and complexity of the brain have kept going upward, at least if we compare points in time that are separated enough for us to see the net effect of frequent upward changes, occasional downward ones, and long periods of quiescence during which not much happens.

But what of the mapping between brain size and behavior? The fact that the former changes gradually does not mean that the latter does likewise. There is an analogy here with traffic. As the number of cars using a stretch of road increases, the road becomes gradually more congested, but for a while the traffic still flows. However, there comes a point where the whole system undergoes a fundamental shift in its behavior, and everything grinds to a halt.

This shift can be described as a threshold effect—but in the case of traffic the threshold is a negative one because crossing it has an adverse consequence for function. The evolutionary buildup of cells in the human brain seems to have resulted in a different sort of threshold—a positive one—the crossing of which has opened up whole new vistas in terms of thought and behavior. We can think about our own evolution. As far as we know, no other creature can perform this mental feat, though if we could communicate better with chimps or dolphins, we might get a surprise in this respect. Anyhow, whatever those wonderful creatures are thinking about, they certainly can't write or read about their own evolution. We clearly *can* do so about ours; so here goes.

CHAPTER SEVENTEEN

CROSSING THE THRESHOLD

There is much to be said for that old adage, "Actions speak louder than words." But in this particular case—the evolution of uniquely human abilities—we should perhaps rephrase it and say that both actions and words speak volumes. Our abilities in both the behavioral and the linguistic domains make it clear that we have truly passed through some kind of threshold in comparison to other animals. Books exemplify both domains: they are wordy products of an intellectual, social, and technological milieu.

WHEN Jack climbed the legendary bean stalk, he passed through a threshold of clouds from the terrestrial domain of everyday life to an aerial one with a magical castle inhabited by an evil ogre. This crossing-through-the-clouds provides us with an interesting, though imperfect, metaphor for the threshold that humans crossed when our upward-evolving brain size and complexity took us from one domain to another: from the one in which most animals are anchored to one characterized by language and abstract thought.

The imperfection in the metaphor relates to the cannibalistic nature of the boy-eating ogre. The evolutionary domain in which humans have arrived could be said to have an evil ogre for sure—our capability for genocide far surpasses anything the lower animals do. But our capacity for good and our potential for great things is also a feature of this new domain in which we find ourselves: medical science and space exploration spring to mind as worthy examples. So perhaps we could modify the bean-stalk metaphor and think of ourselves as having climbed life's ladder into a domain where there is a wise guru as well as an ogre.

Our crossing of the threshold happened very recently. Let's think in terms of orders of magnitude of evolutionary time. The world began a little less than 5 billion years ago; the Cambrian explosion, which among other things produced the first vertebrates, took place a bit before 500 million years ago; by 50 million years ago the dinosaurs had gone and the mammals had diversified to produce many different forms, including primates—monkeys and such. But what I want to focus on is the period that began a little more than 5 million years ago—the last tenth of a percent of terrestrial time.

It was about five to seven million years ago that our own evolutionary lineage split off from that of our closest living relative, the chimpanzee. This fact should be interpreted as meaning not that present-day chimps are our ancestors—clearly they are not—but rather that we and they had a common ancestor way back then; and it looks like the chimp lineage has changed much less since the split than our own.

Much research effort has been focused on what has happened to the human lineage over the previous five million or so years. And views on this have changed considerably over the last few decades. There have been numerous fossil finds of previously unknown species of protohumans, and the prevailing picture in this area of research has shifted from a simple twig with few branches to a vigorous bush with many branches.

The twig view that prevailed when I was a student was popularized in a book I read at the time—*African Genesis*, by Robert Ardrey.

This made much of the carnivorous humanoid "southern apes" that pursued their prey wielding antelope thigh bones as clubs. Human evolution was seen pretty much as a ladder from them to us, with occasional side branches that led only to extinction—such as the Neanderthals. This ladder of progress was beautifully captured in Stanley Kubrick's movie masterpiece *2001: A Space Odyssey* when one of the southern apes threw his bone-club into the air, where it turned into a spacecraft, thus condensing the whole evolution of toolmaking into a few seconds of film.

I said in an earlier chapter that evolution defied simple analogies—that it was part tree, part bush, part ladder. The fact that we now recognize many more species of humanoids than were known back in the 1960s and '70s means that our mental picture of human evolution has become more complicated and bushlike, but this does not mean the ladder has disappeared. What I will do, in this chapter, is accept that there have been many branches en route from our parting of the ways with chimps to our current privileged position; but I will ignore all those side branches that led to extinctions and concentrate on the single lineage that ended up with us. There is no doubting that this lineage climbed a very important rung of life's ladder.

The main changes that took place in this lineage were the adoption of an upright posture of walking on two legs (bipedalism); the pronounced reduction in the amount of body hair; and, most important of all, the expansion in the size of the brain, and of certain parts of it in particular. As you might imagine, I'm going to focus on the last of these changes and its immense consequences.

Let's digress for a moment to put these changes into their geographic context. Our humanoid ancestors are called southern apes because these creatures all lived in sub-Saharan Africa. This particular half of this particular continent truly seems to have been the cradle of human evolution. As various humanoid lineages diverged and went their separate ways, most of them remained in Africa. It was only in relatively recent times that escapes from the confines of a single continent took place. Our own species spread through the Middle East and from there to the rest of the world as recently as about 150,000 years ago. This pattern of migration used to be referred to as the "out of Africa" theory. Nowadays it is probably well enough established that it is no longer just a theory. And indeed it looks like there were earlier exoduses from Africa by other humanoid lineages, albeit these later perished. I recently read an article in a scientific journal titled "Out of Africa Again and Again," which nicely summarizes the prevalent modern view of what happened.

Of course, what matters most is neither the precise place nor the precise time but rather the precise changes that took place in the creatures concerned. Let's now look at the changes in brain size, and their behavioral ramifications.

There's a little problem here. Brains generally do not fossilize. They are made of soft tissue, and they rapidly decompose after death, leaving us with no direct trace of their intricate structural details. But luckily, there is an indirect trace in the form of fossilized skulls. The main part of the skull, the cranium, can be regarded as a tightly fitting container for the brain. Thus the volume of space inside a fossilized cranium can be taken as a good estimate of the size of the brain that used to be housed there. Unfortunately, complete undamaged fossil skulls are rare, but at least we can reconstruct them from fragments, and can then measure brain size indirectly from the size of the reconstructed cranium.

Brain sizes, like car engine sizes, are measured in liters. Small cars have engine sizes of around 1 or 1.5 liters; large ones can extend up as far as 3-, 4-, or even 5-liter gas-guzzlers. Big brains like our own are about the same size as small engines. But the brains that our ape cousins the chimp, gorilla, and orangutan possess are considerably smaller—less than half a liter. So as the lineage leading to modern humans progressed through the last five million years, the brain sizes of the creatures concerned approximately trebled. And since body size decreased in this lineage, the increase in brain size was even more remarkable.

Although the exact pattern of increase, and in particular whether it was gradual, stepwise, or somewhere between the two, is not known, it is clear that the biggest part of the increase occurred in the last two million years. Between the human/chimp divergence more than five million years ago and this two-million-year marker, brain size had increased from less than half a liter to a little under three-quarters. So it increased by something like 300 cubic centimeters (cc). But between two million years ago and now it approximately doubled, adding on more like 700 cc. Of course, these figures can be interpreted in more than one way. If we think in proportional rather than absolute terms, going from, say, half a liter to one liter is the same as going from one to two, as both represent doublings.

The unraveling of the story of human brain size evolution has been—and is—a remarkable example of detective work. Lots of clues have been found. They have been pieced together. And a story has emerged from them.

This story can be thought of as a spectrum of tales with earthy and abstract ends. So far, we have been anchored in the earthy: those fragments of skull that can be reconstructed into whole skulls, whose dimensions can be measured, thus giving us a good estimate of the size of long-gone brains. But the more interesting end of the story is the abstract one. So now, having gone from fragment to skull to brain, we need to go further—to brain complexity and function, and so to language, thought, and consciousness.

The part of the brain that performs these higher functions is the cerebrum. It has become especially expanded in humans and is doubtless responsible for our unusually sentient nature. We could perhaps use the adjective "sapient" (meaning "wise," and incorporated into our species' official name of *Homo sapiens*), but whether the sum total of human behavior merits the "wise" descriptor is a moot point.

The human cerebrum is made up of a vast number of cells—somewhere in the billions. And it is characterized by an enormous number of neural connections—links between one brain cell and another. This incredible complexity of structure is the physical basis of our amazing behavioral repertoire.

Insights into the link between brain structure and function come from a variety of sources. One is the study of behavioral impairment in those individuals who have been unlucky enough to suffer brain damage in a car crash or other serious accident. Another is the study of the behavioral capability of chimps. I particularly like this latter, comparative approach. So do many other people, as testified by the popularity of TV programs on the subject. I have watched lots of these, and been absolutely fascinated.

It is apparent from many excellent studies on human-reared chimps that these creatures have mental capabilities that are not obvious from study of their behavior in the wild. Chimps can learn to do lots of things when presented with the right kind of stimuli by a human mentor. They can distinguish between many different symbols, and can link them with real objects. They can converse up to a point. But their potential for learning and communication always falls far short of the equivalent potential in humans. This contrast is important because it takes the nature of an experiment and thus reveals very clearly the intrinsic dependence of our advanced human mental abilities on our unique human brains.

The distinguishing feature of experiments—the thing that sets them apart from other forms of investigation—is that they vary one thing while keeping all others constant. Achieving the latter is, in most cases, more easily said than done, so most experiments fall somewhat short of the ideal one. And this shortfall is certainly true of experiments on the behavior of human-reared chimps. Nevertheless, the important thing is this: the chimps concerned are getting the best possible schooling. They have an intelligent and highly motivated teacher. They often have one-to-one tuition, something that few human schools or universities can afford to do much of. So the difference between the outcomes, in terms of what chimps can learn and what humans can learn, is not a consequence of inferior teaching; rather, it is a consequence of an inferior brain that is responding to the teaching.

All this talk of limited learning capacity sounds rather negative, and I almost feel that I should be apologizing to chimps for my slur on their character—well, on their mental ability, really, which is not quite the same thing. In all those chimp-behavior TV programs the emphasis is positive; the focus is on the amazing things that chimps can do rather than on their limitations. I have no problem in rejoicing, along with others, in the revelations about chimps' previously unrealized learning capability. But for the point I am making here, the ultimately limited extent of this capability is more relevant.

It is clear, then, that human superiority in the use of tools and language is inherently linked to our big, complex brains. There is a world of difference between using a club and using a computer. And equally, the gulf between the limited communication capability of chimps and a Shakespearean play speaks for itself.

But what of consciousness? This is perhaps the most mysterious and yet important human characteristic. Consciousness is our very being. Ironic, then, that we understand so little about it. I remember reading, a long time ago, the book *The Mechanism of Mind* by Edward de Bono, who came up with the concept of "lateral thinking." He explained almost everything about the brain that he covered in the book incredibly well, and in simple terms that a layperson could follow. But with one exception: consciousness. I recall my disappointment with his use of a mirror analogy where advanced brains incorporate a sort of mirror so they can see themselves. In stark contrast to the rest of his book, this analogy seemed to explain nothing at all.

There is much current excitement in the scientific community about the possibility that research on consciousness over the next couple of decades will reveal important new information. I hope it does, but I remain skeptical. I suspect this very thing that makes us what we are will prove to be a tough nut to crack.

The one thing that is already clear, however, is that consciousness is not an all-or-nothing phenomenon. Anyone involved in those chimp studies is likely to say that chimps are conscious beings. And anyone who has spent much time with a canine companion such as a Labrador will find it hard to imagine that a dog has no consciousness. It gets much harder to know about consciousness in lower animals. I doubt if animals that lack brains, like the sponges and jellyfish we focused on a couple of chapters ago, are conscious in any meaningful sense of the term. And a poorly cephalized creature like a centipede, which can function almost normally when it loses its head, would seem unlikely to have any significant level of consciousness. But the idea that an octopus lacks consciousness I would find hard to swallow.

So one way to look at the evolution of the animal kingdom is as a treelike process in which some of the branches evolved big brains, and with them consciousness. It appears that in the context of particular animals with particular body plans making their living in particular ways in particular environments, natural selection favored the kind of behavior that big brains made possible; and there was sufficient variation in brain size in the creatures concerned for natural selection to act upon.

I have no intention of speculating about exactly why big brains were favored in, for example, octopuses, apes, and humans. Such speculation is often written off, these days, as mere "adaptive storytelling" that is untestable and in the realm of so-called hand-waving arguments. If you want to speculate about such things, you can do so without my help.

Now we need to turn our attention from evolution to development. It is always easy, in evolutionary writing and thinking, to fall into that old trap of thinking only, or largely, about adults. In fact, I fell into it a few pages back when I described humans as creatures that have brain sizes in the one- to one-and-a-half-liter range, because this is, of course, only true of humans in their later stages of development.

We saw earlier that the evidence available on the evolution of human brain size is fragmentary and does not allow a conclusion to be drawn about the extent to which the process was gradual, stepwise, or intermediate. My guess is that it was intermediate in that periods of slow evolutionary increase in brain size were punctuated by bursts of more rapid increase; but "guess" is indeed the operative word here.

In development, we are better by far supplied with evidence. We can describe in detail the growth of the brain through the nine months spent in the womb and in the following period of years until growth stops in the late teens. But that does not mean that we can be equally accurate in describing the developmental increase in consciousness. I suppose that in thinking back through our own earlier lives to that point around the age of three or four beyond which none of us can remember anything, we get some understanding of how our consciousness has grown. But that still leaves the period between the earliest embryonic stage with a rudimentary brain and the three-year-old as very much of a black box. This is especially true of the pre-birth portion of this period, because, although we cannot think our way back to our own days as one-year-olds, we can observe infants of this age and draw inferences, albeit perhaps imperfect ones, about their level of consciousness.

Our lack of understanding of the state of consciousness of human embryos at various stages of their development is problematic in relation to the emotive moral issue of abortion. But that is a whole other topic, and one better dealt with by a specialist in medical ethics than by an invertebrate zoologist like myself. I am no more qualified to make comments on this sensitive issue than the proverbial "person on the street."

The fact that humans have large brains, which they have arrived at by the long-term process of evolution and its sped-up counterpart development, is beginning to have major consequences not just for our own species but for many. We have come a long way from our humble beginnings in sub-Saharan Africa, where our small population and limited toolmaking ability restricted our ecological effects to our immediate environment. Today, our billions-strong population and our advanced technology mean that we affect not just a few African forests but the entire planet.

These planetwide effects are much in the news, especially, at present, global warming. But the ecological effect of modern human activities that is most harmful to other species is habitat destruction. If you consult those front pages of an atlas that give specialist maps, one that you will normally encounter is a map of world vegetation types. If you look at this map closely, you will notice that most of the land area of both Europe and the United States is colored in bright green, and a glance at the map's legend will show you that this color represents deciduous forest.

The chances are that if you look out of the window right now, you will see little or no forest. That's because in most places forests have been replaced with farm landscapes and cities. Our planetwide destruction of natural habitats is doubtless causing the extinction of many species, though we have no accurate estimate of how many. Perhaps, especially given the considerable destruction of those beautiful and immensely species-rich tropical rain forests, we are sending enough species to their doom to refer to this process as a mass extinction.

However, we are not alone in causing mass extinctions. There have been several previous ones in the history of life, caused by natural events rather than by the destructive activities of humans—though perhaps our own activities should also be thought of as natural events, since we are ourselves the product of natural processes. Anyhow, putting such philosophical issues to one side, we can say that mass extinctions have been important contributors to the course of evolution. It is now time to look at them in more detail.

CHAPTER EIGHTEEN

DINOSAUR BLUES

Evolution can be thought of as an interplay between creative and destructive processes or events, though the line of separation between these two players is not always clear. A mutation almost by definition creates something new, in the form of a new DNA sequence, but in doing so it may, like the mutation in the hemoglobin gene that we looked at earlier, destroy some vital function, too. Here we focus not on small-scale events like mutations, but rather on planet-scale events, namely mass extinctions. These events are, of course, primarily destructive, yet they may have creative aftereffects.

IN our grand tour of ascents in complexity that has taken up the past several chapters, we have focused on what is *special* about this aspect of the evolutionary process. But we should never forget that these ascents, whether from one cell to many or from headless to headed, also have something *in common* with all those other evolutionary changes that leave complexity unaffected. Like the famous peppered moths that went black when their environment became soot-covered in the industrial revolution, the driving forces were Darwinian natural selection—survival of the fittest—and the supply of variants (ultimately by mutation) upon which natural selection could act.

When I first learned about the peppered-moth story, decades ago, I was immersed in my studies as a biology undergraduate. I was taking a fairly conventional course in that subject, in which evolution was, naturally, central. My fellow students and I approached evolutionary issues from the viewpoint of established biological disciplines, notably genetics and ecology. So we learned about which genes and which environmental factors were involved in a particular process, whether industrial melanism or something else. And in all this we were wedded to the idea of survival of the fittest.

I was, at the time, blissfully ignorant of the fact that there was a whole other perspective on evolution, which was taken by students of long-dead creatures—fossils—as opposed to the living ones that biologists normally study. These students followed their paleontologist teachers in taking the long view of life. To them, a hundred years was an imperceptible period; *real* evolution happened over millions of years. And in addition to this shift of focus, there was another, related one: from looking at minor changes within species (microevolution) to looking at the birth and death of species (macroevolution).

The death of a species we refer to as an extinction. Over the whole of evolutionary time, there have been countless extinctions. Indeed, it is widely accepted that the vast majority of species that have graced our planet with their presence are now extinct, leaving only a tiny fraction still alive and well—or extant, as we call it.

In a typical million-year period, a very small proportion of species become extinct. This is the so-called background extinction rate. But the earth has experienced at least five mass extinction events, in which the number of species disappearing has far surpassed that which would be expected on the basis of the background rate alone. The most famous of these occurred around sixty-five million years ago, when it dealt its deathblow to the dinosaurs, and to many of their contemporaries, too.

To describe how species fare as the earth passes through a mass extinction period, some paleontologists have coined the lovely phrase "survival of the luckiest." To see how this expression relates to survival of the fittest, think again of those evolving peppered moths. If, during the industrial revolution, the earth had been plunged into darkness for several years, for example by asteroid impact, the trees on which these moths rested may well have become extinct; so may the moths themselves and the birds that ate them. Essentially, the whole system would have vanished, and the sort of evolutionary fine-tuning that was going on in the form of the adaptation of the moths' color to the changing color of tree bark would have been overwhelmed by, and replaced with, a much-bigger-scale evolutionary process to which minor variations in the species concerned were irrelevant.

But perhaps, in whichever corner of the world was most distant from the putative asteroid impact, the dust cloud was thin and there remained some intermittent sunlight. Here trees, moths, and birds might all have survived, although, being so far away from the ones we started with, they are likely to have been entirely different species. Many years later, when the climate regained a degree of normality, these "lucky" species may have spread across the world and ultimately populated the areas most devastated by the mass extinction.

There has been a huge growth of interest in these catastrophic processes over the last couple of decades. Most of this interest has centered on the severity of the extinctions—what proportion of species perished—and their cause. I will examine these issues first. But then I will return to our main theme and ask what effect mass extinctions have on the evolution of complexity.

In order to get a grip on mass extinctions, we need to know a little of the history of the earth. But "little" is an important word here; there is certainly no need to delve into the geological past in soporific detail, nor to learn more than three or four of the hundred or so names that exist for various slices of times past. A single paragraph will suffice.

The most crucial marker in geological time, at least from an animal evolution perspective, is the base of the Cambrian period some 550 million years ago. This marker divides the earth's past in two: first, a long early period (about 4 billion years) from which animal fossils are nonexistent or, as we approach the Cambrian, of debatable interpretation; and second, a much shorter, more recent period (just over half a billion years, of course) from which animal fossils are abundant. This latter span of time is conventionally split into three eras: the Paleozoic ("old animals"; 550 to 250 million years ago), the Mesozoic ("middle animals"; 250 to 65 million years ago), and the Cenozoic ("recent animals"; 65 million years ago to the present).

Why such an asymmetrical division? If you intend to split the history of anything at all into early, middle, and late sections, you would probably want some compelling reason to do something other than a simple division into three equal parts. Well, in relation to the history of life on planet Earth, that compelling reason is provided by the very mass extinctions that are the subject of this chapter.

With regard to that most recent and most famous mass extinction, the one that separates the Mesozoic and Cenozoic eras, it has been estimated that about 90 percent of all species on Earth perished. The dinosaurs were the most conspicuous of the animals that disappeared, but in terms of overall species numbers they were just the tip of a very large iceberg. The earlier mass extinction that separated the Paleozoic and Mesozoic eras seems to have been even more drastic; among the animals that were finally sent into oblivion at that time were the trilobites, which are among the most famous of all extinct invertebrate groups.

Most accounts of the geological history of the earth acknowledge about three further mass extinctions in addition to the two classic ones noted above. And it may be that the earth is entering yet another phase of mass extinction, but one that is unique in that its main causal agent is us—as we noted in the last chapter. Human activities are having planetwide effects, as is well known to almost everyone through articles in the popular press about global warming, holes in the ozone layer, and habitat destruction.

But this book is not the place to chronicle further details of multiple mass extinctions. Rather, I now want to look briefly at possible causes, and then move on to effects on complexity.

Debate has raged most fiercely over what caused the demise of the dinosaurs and many of their contemporaries, so let's start with that particular mass extinction, the one that took place sixty-five million years ago. There are two main competing theories as to how it may have come about. These are asteroid impact and widespread volcanic activity. The former has had more hype, but that does not necessarily mean it is more likely to be right.

One of the most compelling lines of evidence for an asteroid colliding with the earth about sixty-five million years ago is a high concentration, in rocks of that age, of an element called iridium. This is a member of the platinum family of elements, and, as the price of platinum testifies, these elements are generally rare on Earth. However, iridium in particular is known, from chemical analysis of meteorites reaching Earth from the asteroid belt, to be much commoner there than here. So if an asteroid of a considerable size (miles across) hits Earth, disintegrates, and sends up a cloud of particles into our atmosphere, what we would expect, when the dust settles, is a layer of unusually high iridium concentration. And this is precisely what we see in rocks from sixty-five million years ago.

If we were only to consider this single source of evidence, it would seem like an open-and-shut case. But, as ever, history defies simple explanations. To see why simplicity does not prevail in the case of the Mesozoic/Cenozoic mass extinction in particular, we need to take a trip to India.

The quasi-triangular Indian subcontinent is something we can all picture in our mind's eye, whether from high-school geography lessons, casual perusal of an atlas, or pictures in the newspapers. As you might expect for such a large piece of land, its geology is rather varied. Most of it I will quietly ignore, in order to focus on just one thing: a vast expanse of the rock type called basalt in the west of the subcontinent.

Basalt is an igneous rock; in other words, unlike those rock types (for example, limestone) that form through accretion and compression of fine particles, basalt originates from volcanic eruption. Extensive areas of basalt in the present thus point to extensive volcanic activity in the past. Now, if you look at the geology of western India, what you find is an area of basalt covering many thousands of square miles, from Bombay on the west coast to Nagpur in the center of the subcontinent.

How old is this vast expanse of basalt? Well, you've probably guessed it by now—sixty-five million years. We know this age not just from dating the rocks but also because the iridium layer actually runs through them. So we have a second suspect. Extensive volcanism could cause earthwide dust clouds just as could an asteroid impact. Which was it?

Let's leave that question hanging for a moment and turn to the earlier mass extinction that occurred about 250 million years ago. Is there a layer of very high iridium concentration in rocks of that age? Well, actually, no. Are there extensive areas of basalt suggesting massive volcanic activity at that time? Intriguingly, yes. There is an area of basalt stretching across northwestern Siberia of approximately equal area, again many thousands of square miles, to its Indian counterpart. It coincides exactly with the mass extinction that took place at the end of the Paleozoic era.

The most economical, or parsimonious, hypothesis for the cause of mass extinctions is thus volcanic activity, not asteroid impact. But if this hypothesis is correct, it leaves us with a problem. Was the asteroid impact that almost certainly took place at sixty-five million years ago just a coincidence? In other words, was it something that happened while those countless volcanoes were erupting and polluting the air with poisonous gases and dust? Perhaps, in doing so, it added its own destructive effects, thus rendering this mass extinction especially severe. Or alternatively, was it not coincidence at all?

Lurking behind this last question is the possibility that the sixty-five-million-year-ago collision actually *caused* widespread volcanic activity as its shock wave fractured the earth's crust. If you want to learn about the likelihood of such a connection, it is time to put this book aside for a moment and consult a popular geology book written by someone better equipped than I am to deal with such issues.

And so, finally, to the effect of mass extinctions on the complexity of creatures. To consider such effects, we need to build up a mental picture of the world as it was just before a mass extinction struck. It is probably easiest to do this for the extinction that ended the Mesozoic era, because we have all seen TV documentaries about the dinosaur-dominated world of that time. But we also need to have two other mental pictures to compare with this one: the world immediately after the mass extinction; and the world as it has become some ten million years later, after evolution has had sufficient time to make good the gaps in the fauna opened up by the volcanoes and/or asteroid.

Let's look at these three points in time. Also, let's concentrate for the moment on the vertebrate fauna of the land to avoid our mental pictures becoming too complicated.

Just before sixty-five million years ago, numerous types of dinosaurs roamed just about every corner of the terrestrial world. (That world, incidentally, looked rather different from today's. For example, North and South America were separated by a large ocean, as were India and the rest of Asia.) But what of other groups of land vertebrates?

At that time, there were, in addition to dinosaurs, many other well-known reptile groups, such as crocodiles and lizards; also amphibians, birds, and mammals. So all the main groups of today's land vertebrate fauna were already in existence. Thus the mass extinction event, dramatic though it was, did not take the form of a series of origins of new groups to in some sense replace the disappeared dinosaurs. Rather, it took the form of proliferation (or radiation, as it tends to be called) of already-existing groups.

Such radiations take time. So, immediately post-dinosaurs, nothing much happened. But several million years later, there had been massive radiation of the mammals, and also of the birds.

How did these events—extinctions followed by radiations—affect the average complexity of land vertebrates? This is a tricky question because we are focusing here on animals that are all very complex when considered against the general background of the whole animal kingdom. The differences between them are small compared, for example, with the difference between any of them and a really primitive animal such as a jellyfish or a sponge.

I'm going to concentrate on behavioral complexity and, associated with this, complexity of the brain, especially its main thinking part, the cerebrum (or cerebral hemispheres). Compared with the rest of the brain, and also with body size, the cerebrum is generally largest in mammals, smaller in birds, and smaller still in reptiles. Larger cerebrums allow more different kinds of neural interconnections and hence permit more complex patterns of behavior, including greater learning ability. Think about the learning ability of chimps (excellent) and garden birds (surprisingly good) as seen in nature programs on TV. And notice that few such programs even attempt to show lizards performing learned tricks.

Given these comparisons, it's clear that the eventual replacement of the dinosaurs by a terrestrial vertebrate fauna dominated by birds and mammals was a step up the ladder of average complexity. Of course, as ever, individual lineages may have been bucking the general trend. Increases in complexity are not inevitable. There is no universal law of evolutionary increase in complexity that applies to every bunch of creatures. Recall that parasites, for example, often evolve downward in this respect. We do not want to end up back in the dark days of the nature philosophers, with their notion of a neat vertical line of creatures with humans at the top. Evolution is part lawn, part bush, part tree, part ladder. It defies simple models.

Before leaving our vertebrate focus behind and spreading out to consider the whole of the animal kingdom, I want to dispel a couple of popular misconceptions about birds and mammals. First, these two groups are not very closely related. They had quite separate origins among different groups of reptiles. They have both evolved warm-bloodedness, that is, the ability to maintain a constant, and quite high, internal temperature, despite the vagaries of the external environment. But they have done so independently. This is a case of convergent evolution, not inheritance of a shared feature from a common ancestor.

Second, the origin of birds was considerably *later* than that of mammals—the opposite of what is popularly thought. Everyone is familiar with the middle period of the Mesozoic era—the Jurassic—if only from the movie *Jurassic Park*. The first mammals were already in existence at the start of the Jurassic, but the first birds did not appear until near its end. Since the Jurassic lasted for about fifty million years, there is a considerable temporal gap between the origins of the two most advanced groups of vertebrates.

But the story so far is, in a sense, very parochial. Vertebrates may indeed be the most complex of all creatures, but they are nevertheless just a tiny corner of the living world when considered in terms of number of species. They represent about 5 percent of known animal species today. Furthermore, given that our knowledge of most things living is very vertebrate-biased, they probably represent much less than 5 percent of *all* species—that is, the number we would get if we were able to add in all those extant species that are as yet undiscovered and unnamed.

So what about the effect of a mass extinction on organismic complexity more generally, outside the confines of one particular animal group? Recall that along with the dinosaurs, many other animals, including countless species of marine invertebrates, met their end. Others survived and prospered. Some radiated. In previous mass extinctions, such as the one at the end of the Paleozoic era, there were no photogenic dinosaurs to distract our attention from these broader-scale effects.

I can't help but feel that mass extinctions are by their very nature unpredictable events in more ways than one. I don't think we can generalize about their effects on the average complexity of creatures. Sometimes, as in the case of the sixty-five-million-year-ago extinction, the long-term effect may include the radiation of a particularly complex group. But this outcome is far from inevitable. The ascent of life's ladders by certain lineages, and by the "average creature" inasmuch as this is a meaningful concept, has been a consequence of the accidental invention, through mutation and duplication of genes, of the very components of which complexity is made. These components include that supreme building block the eukaryote cell, the Velcro that enables such cells to stick together, organs composed of different cell types, and the big brains that lead to complex behavior.

These components of complexity all emerge from the creative side of the evolutionary process, in ways that we have examined over the course of the last few chapters. The destructive side of evolution, in which mass extinctions are the most impressive players, may sometimes facilitate its creative counterpart by clearing the way for complexification. But that's all. The rising complexity of creatures is a consequence more of small-scale accidents at the level of the gene than of large-scale accidents at the level of the biosphere.

CHAPTER NINETEEN

BEYOND PLUTO

I strongly believe that there are no life-forms of any kind on any of the other planets in our solar system. Within this spatial constraint, I believe that we are indeed alone. But if we think (or ultimately travel) further afield, I believe equally strongly that we are not alone—that there is life, probably even intelligent life, lurking in at least one other quiet corner of the universe. If so, what is it like? And should we expect it to be characterized by an evolutionary increase in the average complexity of its creatures over time, paralleling that found here on Earth? We probably should.

IN the previous chapter we delved into Earth's history. Having history in the picture urges a certain type of question. What if a dinosaur-killing asteroid had started its sunward trajectory at a very slightly different angle, so that, from a terrestrial perspective, it was a near miss, rather than a hit? Or, if you incline toward the alternative theory, what if an outburst of volcanic activity sixty-five million years ago had never happened? Would the dinosaurs (and countless other extinct creatures) have survived and prospered? Would mammals have remained eclipsed by their gigantic reptile cousins, and never progressed from their humble beginnings to their present grandeur?

These questions are the beginning of a slippery slope, albeit to a productive place: the mental state of not taking things for granted.

There is no reason to restrict our questions to just one of the five or so mass extinctions. Moreover, there is no reason to restrict them to mass extinctions in the first place.

Consider smaller-scale evolutionary events. Each of the great groups of creatures, whether vertebrates, insects, mollusks, or whatever, had its origin in a particular place at a particular time. What if conditions there / then had been different? Might the stem that grew and radiated over the eons never have appeared? Might it just have taken some small ecological difference in some remote corner of the world for vertebrates to have never arisen? Or, taking a comparable "internalist" stance, might the origin of a major group of animals never have happened if some highly improbable variant developmental trajectory had not been brought into existence by mutation, and thus had no chance to take root more widely through natural selection? (The turtle shell springs to mind as an example here.)

We have encountered, at many stages of this book, that most fertile of recent thinkers about evolution, the late Stephen Jay Gould. And now we do so again, for he came up with a brilliant metaphor for these questions that covered them all at once. He asked: What would happen if we were able to replay the tape of life? That is, if it were possible to wind life's videotape back to "year zero" sometime between three and four billion years ago, press STOP, erase everything on the tape, take a deep breath, and then press RECORD, and later PLAY, what would mesmerized observers of this unique experiment see?

Perhaps the living world that began to unfold would be similar to the one we know, right down to the individual species. At the other end of the spectrum of possibilities, perhaps the cast of characters on the replayed tape would be unrecognizable because of some critical early difference—maybe all remaining unicellular because none of the cell types that emerged was able to produce the glue necessary for multicellularity. In between, there is a multitude of options. These include the one that we started with, in which the dinosaurs' reign remained unbroken sixty-five million years ago and humans languished in the realm of possible but unrealized creatures.

Space travel is possible (albeit very limited to date); time travel is not (well, at least as far as we know). We will not be able to investigate these fascinating questions about life's repeatability by going back and changing something to produce a "fork" in the temporal progression of some particular event, as the time technicians did in one of Isaac Asimov's science fiction novels. But perhaps it will be possible to investigate questions about life's repeatability through space travel. Suppose we can find a planet very much like Earth. Suppose further that we can date it accurately (quite probable) so that we compare like with like—for example, a half-billion-years-younger planet could be compared with Earth at its immediately post-Cambrian-explosion stage. Then we can get at life's repeatability after all.

There are, of course, some problems, and to say that they are big ones would be something of an understatement. The two most important of them are as follows: (1) finding, and getting to, a suitable planet; and (2) the difference between suitable and identical. I'll take these two problems in turn.

I don't doubt for a moment that exceedingly Earth-like planets exist. But they certainly cannot be found in our own particular neck of the cosmic woods. Our solar system has at least eleven planets, and there are probably still other far-flung ones awaiting discovery. Regardless of the precise number, let me make a bold prediction: However many space probes are sent out to Mars, Titan, or elsewhere in our solar system, none of them will find complex creatures. In fact, to be bolder still, none of them will find life.

These predictions reflect my long-held view that outside of our own planet, the solar system in which we are located is a barren, lifeless place. The range of conditions under which life can evolve is very narrow when compared with the range of conditions that can be found. For example, in terms of temperature, life probably needs to evolve in the range of 0 to 50 degrees centigrade, whereas in deep space the minimum is not our fake zero, but the "absolute zero" of minus 273 centigrade that represents a total absence of heat, and in the center of the sun the temperature reaches millions of degrees. While the various planets in our solar system do not reach such extremes, they nevertheless all have temperature ranges that wander way outside of those that are bearable by life as we know it.

But do we really know it? There are tiny creatures called water bears, just a millimeter or so long, that are distant relatives of the insects. You can find them by rinsing mosses, catching the runoff water, and inspecting it under a microscope. It turns out that these curious creatures can be taken down to a temperature close to absolute zero, be warmed up again to room temperature, and then walk off on their eight stumpy legs, apparently none the worse for their deep-frozen sleep. On the other hand, it is one thing for a creature to *survive* a temperature of less than minus 200, quite another for it to *evolve* at that temperature in the first place.

There is another dimension altogether of this question about life "as we know it." Perhaps life beyond Earth can take myriad forms, including some entirely unexpected by a terrestrial thinker. Perhaps life-forms can be gaseous and exist as living clouds in the atmosphere. Perhaps some are based on a different element than carbon (the basis of all creatures on Earth), and so have a weird (to us) biochemistry allowing them to exist miles from an airborne or waterborne source of oxygen. There is no reason why such life-forms may not be found near their planet's core rather than stuck to its surface by gravity.

It's time for another bold prediction. I'm going to assert that there are no such bizarre life-forms anywhere. Given the vastness of the universe, such a prediction may, in the long term, be doomed to being proven wrong. But I'll stick with it for now, if only because there is, as yet, no evidence for life of these kinds, and it's better not to let our imagination run completely wild. Treat this as a wielding of Occam's razor, if you like.

Let's assume, then, that there is no life in our own solar system outside of planet Earth: no gas-creatures lurking in the "gas giants" Jupiter and Saturn; no core-creatures hidden near the center of Venus or Mars. If this is true, then to look for life we need to travel light-years beyond Pluto.

A reminder of the broad structure of the universe may not go amiss at this point. Beyond Pluto (and the newly discovered outer planets) there is a massive stretch of space before we encounter the next solar system. The distance is so great that a space probe heading out that way, which took ten years to get to Jupiter, will take not tens but millions of years to reach Solar System 2. The universe consists of a series of galaxies (star clusters); each galaxy probably possesses numerous solar systems, though we don't know how many, because it is not yet clear what proportion of stars (that is, suns) have planets.

This seems to have become the chapter for sticking my neck out, so here are a few more bold predictions. There are lots of solar systems with life, and indeed not just life but complex-creature-life. All feature carbon-based biochemistry; all involve bounded batches of semifluid living matter (cells); many have evolved multicellularity. The key question now becomes: What will those complex multicellular creatures out there—say, in Solar System 42 somewhere in the great galaxy of Andromeda—look like?

How do we get to that closest of all galaxies, in Andromeda, that is visible to the naked eye of a shrewd nighttime observer yet is light-years away—or even to a remote corner of our own Milky Way galaxy? Traveling at the speeds possible for today's spacecraft, no one setting out on such a journey would live to see the end of it. The possible ways around this problem are suspended animation (like the water bears); a multi-generation venture; and vastly greater speed. The last of these options may be subject to Einstein's limit—that nothing can travel faster than the speed of light—but then again it may not. After all, the apparent impregnability of Newton's physics fell to Einstein's relativity; perhaps that in turn will fall to something else.

Suppose that in one of these ways, the first of our two big problems—that of getting there—is solved. The second big problem then becomes apparent. Not only can we not test life's repeatability by replaying its tape on Earth, but neither can we test it by watching the tape play on some Earth-replicate elsewhere. This is because perfect replication is a myth. Even in a classic lab experiment, where, for example, I set up would-be replicate microcosms of fruitfly life in an incubator, these are never true replicates. One tube of food is never identical to the next. Close perhaps, but never exactly the same. And this lack of perfect replicability that is true of tubes in incubators is even more true of planets in solar systems.

But fear not. The lack of *precise* replication of planetary conditions does not spell the end of our quest. In experiments, the typical approach, given the impossibility of precise replication, is to do not just two but many replicates. In the realm of space we can do likewise. So we would visit many Earth-like planets and observe what form life (if there is any) takes on each.

Back, then, to the key question, and to my final series of bold (but perhaps wrong!) predictions. I suspect that all Earth-like biospheres, if sufficiently long established, possess complex creatures that have climbed life's ladder in evolutionary terms, and continue to climb it every generation in their own individual development. I suspect that extraterrestrial evolution will, in all of our quasi-replicates, be driven by a process of natural selection, based on a limited, and biased, supply of variants.

Even if all these predictions are right, what will we observe on descending from our planetary lander and looking around? Of the creatures that are big enough to see, will they be recognizable as plants and animals? Will we see trees, birds, insects, mice, and, most intriguing of all, humanoids?

Even for someone not averse to coming up with predictions, it is hard to predict the answers to these final questions. This is because we are getting into detail. My guess, for what it's worth, is that broad body plans would be predictable; that plants and animals would be found; that walking, swimming, and flying would all be encountered as ways of getting around, with consequent implications for the body forms of animals moving in these ways. And it seems likely that on many Earth-like planets, intelligence has evolved.

But I doubt if things are predictable at the level of the species, as opposed to that of the body plan. In artistic terms, the broad-brush picture may be predictable but the fine brushstrokes are not. A world in which undersea cities populated by intelligent, book-reading octopuses are unmatched by any land-based civilizations seems entirely possible.

Our perceptions of time and space are inextricably interwoven.

If I make a long journey, say to another continent, the things that happened a few days ago feel longer ago than they should. And, related to this, our concept of simultaneity becomes fuzzier as the distance between simultaneous events lengthens. In other words, the farther away something is from our current location, the harder it is for us to really *feel* that it is going on at exactly the same time as the things we observe around us.

I have no doubt that as I write this, there are tadpoles developing in the pond in my garden. I cannot see them right now, but I know they are there. I can picture them in my mind. Move a bit further afield and the picturing process becomes harder. Just as I write this sentence, and, later, just as you read it, someone is experiencing his or her first kiss, someone is being tortured, someone is being born, someone is dying. But I cannot picture them in sharp focus. Perhaps it is because there are too many of them; perhaps it is because I don't know them personally; but perhaps also, at least in part, it is because they are far away.

This distance-as-time phenomenon becomes especially acute when we are contemplating distant planets. I have had to keep going back over drafts of this chapter, changing the tense of my verbs. I find myself writing things like, "Biosphere 194 would have creatures like . . . ," when what I mean is that it "has." Okay, there is an element of uncertainty about the appearance of the creatures concerned, but the fact is that they are sitting/standing/walking/swimming there *right now*, going about their alien business, just as you and I are going about ours. Perhaps some of them are even reading a book about the nature of life.

What will be our first close encounter with these unseen fellow creatures? When will it occur (in our lifetime?), what form will it take (radio waves or personal visit?), and will its effects be good or bad? Will we go the way of the Native Americans if confronted by a technologically more advanced civilization? Or will our culture be enriched through cross-fertilization of ideas with another? I have run out of predictions; these questions must remain open.

Some scientists hate what they call speculation. This chapter has been full of it. So perhaps a brief defense of its inclusion is warranted. Personally, I spend a lot of time designing experiments, collecting results, analyzing them, and drawing carefully measured conclusions peppered with ifs and buts to make sure I do not go further than is warranted by my findings. Most scientists do likewise. And to be sure, there is little room for speculation in this endeavor, except perhaps to say that experiments are designed to test hypotheses, which are, by another name, speculations.

But in a lifetime of doing particular experiments on particular things, whether the response of insect development to varying population density or the effect of the chemical composition of pond water on the way a snail's shell grows, is there not room for a little speculation about the nature of extraterrestrial life? Would we all not be a little poorer intellectually if we were to fail to speculate about the big issues—those that cannot be addressed by specific experiments? Of course we would be. We *should* speculate about such things. But we should maintain a clear line of demarcation between this fascinating if inconclusive activity and the day-to-day practice of science.

Well, so much for the universe. We are now about to go even further afield. And those dyed-in-the-wool anti-speculationists who have read through the present chapter with mounting concern will be even more aghast at the nature of our final journey (in more senses than one). But here it comes anyhow.

CHAPTER TWENTY

BIG QUESTIONS

What is the relationship between science and religion in relation to the origins of the complexity of creatures? The answer to this question is not a simple one, except in special instances such as the debate between an evolutionist and a biblical literalist. In this case, the former is right, the latter is wrong. But what of less clear-cut debates, such as that between an evolutionist and a proponent of intelligent design? And, in broader terms, should belief in an evolutionary process that is unaffected by the tinkering of a designer necessarily lead to atheism?

HOW do you measure the size of a question? What makes one question bigger than another? And so what urges me to adopt the title that I have for this, the final chapter, especially given that the questions we have been pondering in the previous one—about the nature of life across the universe—were very big indeed?

The questions that follow are bigger still. They are even more vital to our overall worldview. They also concern our own personal futures, in the sense of whether we continue to ascend life's ladder in some as-yet-indiscernible way, after our terrestrial deaths, or merely slide down it into the mud. And, associated with this issue, they concern the question of whether there are dimensions and universes beyond the current four and one, respectively, of which we are aware. Here science meets not just religion but science fiction. I have encountered the idea of multiple universes, or a multiverse for short, in two places recently: a science journal and the film *The One*, starring Jet Li. Also, religious beliefs would appear to suggest extra dimensions or universes, assuming that most Christians, for example, do not believe that heaven is on some remote planet.

To research this chapter, I did something I had never done before: I visited some Web sites representing creationism in its many guises. This exercise was a revelation indeed, but probably not of the sort that the Webmasters had intended. What I found most striking was the appalling lack of integrity of those concerned. The deliberate misuse of quotations and details from the work of scientists suggested that all honor and honesty had been cast to the four winds. I realized that I was in a different social context from the one I have known and loved for my whole scientific career, where an honest search for the truth is at the heart of things. Instead, I was in a milieu where the dominant ethos was to force acceptance of a particular worldview by any means whatever. No holds barred. Not the Spanish Inquisition perhaps, but the intention seemed the same: to stifle freedom of thought. And it mattered not whether I was in the grips of young-earth creationists or intelligent-design proponents. The latter were more slippery and difficult to pin down, but always in the end I found evidence of dishonesty.

Now, if you are a religious liberal, you are probably reading this with a sense of growing concern, even anger. Why should I let what some would regard as a lunatic-fringe movement (however large it may be in certain countries) represent the whole of religion? Is this not a bit like letting a scientist who has been convicted of fraud represent the whole of science? Well, maybe it is, but there is method in my madness. I will get to the liberals in due course. But I want to split this discussion into two parts: one that is firmly universe-bound and one that is not. And I'll take them in that order.

The creationists and their ilk will indeed represent the religious camp in the first part, but I will not let them do so in the second.

Here is the worldview of a typical scientist. (I suppose "universe-view" would be more appropriate, but no such term is in wide circulation.)

Sometime about fourteen billion years ago, there was an event that has become popularly known as the big bang. It was an instantaneous thing. Before that instant there was no matter or energy, and indeed no space or time, either. After that instant, all four existed. In the period from then to now the universe has been expanding. It is still doing so, and we can measure this by a phenomenon affecting light waves called the redshift. This is a bit like the phenomenon with sound waves where we can tell with our eyes closed that the siren of an emergency vehicle has passed us and is moving away, because of the altered pitch of the sound and its pattern of change.

Measurements on celestial bodies show redshifts everywhere. Everything is moving away from everything else. This is exactly what we expect in the aftermath of an explosion, whether the big bang or something smaller. As stars fly apart from each other, some of them develop solar systems, whether by condensation of rings of dust or by other means. Conditions on the different planets vary widely: some are gaseous, others solid. The solid ones take a while to cool from their fiery beginnings. Some (a very small proportion of a very large number) become suitable for life. Earth at about four billion years ago was one of those planets.

Continuing the story for Earth only: Life appears gradually, starting from reproducing aggregates of large molecules in the primordial soup. It becomes cellular, then diversifies wildly, with some lineages of creatures remaining at the bottom rung of the complexity ladders (unicells), others racing up them, to reach bodies consisting of trillions of cells. One of these complex creatures becomes conscious to a higher degree than all others and begins to think, write, and read about its own origins.

So far so good. The last paragraph is really just a rapid recap of this whole book. But there is another side to this overall scientific view of the evolution of the universe in general, and terrestrial life in particular, that has up to now remained implicit. I will now make it explicit and, in doing so, will reveal the magnitude of the gulf between science on the one hand and religious fundamentalism on the other.

Continuing the worldview of the typical scientist: At all stages since the big bang, everything that has happened in the universe has been subject to mathematical, physical, and chemical laws. There have been no divine interventions in the form of major miracles or minor tweakings of the course of events. If there is a divine being, which I regard as an open question and will discuss further below, she has stood back from her creation and let it develop in its own way. (Note that, confronted with the choice of "He," "he," "She," "she," "It," or "it," I have made a quasi-random choice in favor of one of them.)

Many consequences flow from this single short paragraph. If the views espoused in it are true, then there is no point in praying, in the sense of imploring a deity to make certain things happen or not happen, though there may be much sense in praying in that other sense of quiet reflection bordering on meditation. Also, if there have been no miracles, then there was no virgin birth of Jesus, though he may well have existed on Earth as an exemplary man from whom we could all learn much. There is no such thing as a "holy war," a contradiction in terms if ever there was one, whether perpetrated by Christians against Muslims in the days of the Crusades or by Muslims against Jews and Christians (and even against fellow Muslims) in the form of present-day suicide bombers. Such bombers blow apart infants too young to have any religious views, so that their heads go one way, their limbs another, and their blood-spattered torsos another still. I grew up against a background of such slaughter in Belfast, where one bunch of so-called Christians committed these acts on another. Jesus would have been horrified.

What makes the scientist's naturalistic view of the universe any better than its divine-interventionist counterpart? Well, precisely this. Most of those who hold the naturalistic view do so in a tentative way. The English evolutionary biologist John Maynard Smith put it memorably when he said that, unlike a creationist, he was prepared to abandon his stance if he were to find strong evidence against it. He contemplated the possibility of finding a new family of fish in which each species had a different pattern of spots on its tail, and every one of these patterns took the form of a constellation of stars—for example, Orion the Hunter on the tail of one species, the Great Bear on another. He said that if such fish were found, he would have to reconsider his support for evolution.

It is precisely this tentativeness, and a willingness to consider the possibility of evidence contrary to one's worldview, that is lacking from the stance of creationists. Their writings reveal quite the opposite: a false certainty, and a desire to distort any evidence so that it appears to support them.

The Web sites of those slippery intelligent-design people make a big play of arguments based on complexity. They claim (like Archdeacon Paley, two centuries earlier) that the existence of complex creatures argues against a naturalistic, evolution-based origin. This is crass nonsense. The evolutionary processes that I have described, such as duplication and divergence, easily, even elegantly, explain how creatures have climbed life's ladders, unaided by supernatural agencies. We have progressed far indeed in the last two centuries in our understanding of such processes, and the century that lies ahead promises to cement and refine this understanding.

The pursuit of a rational understanding of the world is one of the most noble goals of humankind. Another is the rejection of false certainty, superstition, and all attempts at denying others freedom of speech through forced recantation (Galileo) or murderous fatwa (Rushdie). Goodwill and success to all those involved in these noble endeavors in the continuing battle against the dark forces of fundamentalism.

Well, so much for the first part of the story, the one concerning the battle between naturalist and creationist views of the way things are. All the evidence I have ever seen favors the former. There are no fish with heavenly-constellation tails. And it makes no sense that a divine being interferes in worldly affairs intermittently, capriciously granting prayed entreaties here, denying them there. Now we turn to that other side of the religious view of life, the one that concerns whether "things" exist in domains outside of the single four-dimensional universe that we can observe. These "things" could include the Creator herself, our own immortal souls, if we have them, and anything else you might care to speculate about.

It is here that the going gets tough. I have no hesitation in throwing my weight against the creationist view of the world. But I do hesitate if asked to align myself, on matters beyond the bounds of the universe, with a convinced liberal pro-religionist (whether Christian, Jew, Muslim, or whatever) or an equally convinced atheist like the English biologist Richard Dawkins, author of that beautifully titled book *The Blind Watchmaker*.

My reason for hesitation is simply that there is no evidence either way. My scientific instinct tells me to look for evidence. With regard to life on Earth I find lots of it. And it all argues in favor of evolution. But with regard to realms which, by definition, permit no examination and thus provide no evidence, I find nothing to guide me. I see no more reason for a rational scientist to be a committed atheist than to be a committed theist. I do, in contrast, see every reason for such a person to be a committed agnostic, if that is not a contradiction in itself.

We owe the word "agnostic" to the man who earned the nickname Darwin's bulldog for his fierce defense of evolution against diverse nineteenth-century critics—Thomas Henry Huxley. It is hard to overestimate Huxley's importance in both respects. The initial progress toward widespread acceptance of evolution by natural selection would have been slower by far without Huxley's articulate and impassioned advocacy. And what I regard as the only honest worldview would have continued to be anonymous without the name that Huxley gave it, and so to be less visible beside its faith-based counterparts of theism and atheism.

It might seem odd to categorize atheism as being faith-based. Yet the more I think about it, the less I am able to see it in any other way. No intelligent and honest person could believe, any longer, in a Creator who supposedly made the world and all its creatures some six thousand years ago, as was argued by the seventeenth-century Anglo-Irish archbishop James Ussher. In the twenty-first century, the possible Creator that we can decide to believe in or not, or to remain agnostic about, is an altogether more elusive one. The existence of a Creator inhabiting as-yet-unknown dimensions, whose only involvement with the universe was to launch it forth from nothingness (why?), who does not direct the evolutionary process that eventually started on at least one planet, and who does not reveal herself to humans (unless after death?), is impossible to decide upon with the evidence available to us. So belief in her absence is as much an act of faith as belief in her presence. Agnosticism, on the other hand, represents a lack of belief in either direction. More broadly, it represents a philosophical stance in which we refuse to believe in something (or its lack) without adequate evidence.

This stance is one of the foundations of science. Through it we have come to understand how and when complex creatures arose, and much else besides. The reliance on evidence, and the lack of reliance on faith or authority, have been essential ingredients in the progress of human thought. No book is infallible. This applies to the holy books of all the world's religions, and to all scientific books, including, of course, this one. We owe it to ourselves to reflect upon what we read, and not to take anyone's word as unchallengeable. As Huxley said: "The ultimate court of appeal is observation and experiment, and not authority."

Not only is the progress of human knowledge important, so is the life of every individual. Intolerant faith has often resulted in the taking of lives. We have already discussed the Crusades of almost a thousand years ago and the terrorist bombings that have become an unfortunate feature of our present-day world. These takings of innocent lives were, and are, perpetrated by people who claim allegiance to one or another religious faith. But we should also remember that a mere few decades ago, millions were sent to their deaths in Stalinist Russia, many of them because their religion offended the atheist faith of the Soviet regime. To my knowledge, not a single human life has ever been taken in the name of agnosticism.

GLOSSARY

I give here working definitions of some important words used in the book that may not be familiar to all readers. In some cases, I add examples or other useful information. Within most of the entries, some words are italicized. This means that they also have Glossary entries, and so provide an opportunity to cross-refer where appropriate.

AGNOSTICISM Philosophical stance of those people (agnostics) who admit that it is not possible to know for certain whether God exists or not. Thus distinct both from faith-based religions and from *atheism*. The word was introduced by Thomas Henry Huxley ("Darwin's bulldog") in the late nineteenth century.

AMOEBA Single-celled *creature* illustrative of the level of *complexity* that life-forms had to pass through to get from a bacterial state to a *multicellular* one. The *cell* of an amoeba possesses a more complex internal structure than a bacterial cell, and resembles animal cells in possessing a *nucleus* that contains the genes.

ANTERIOR The head end of an animal; at or near the head end. The opposite end is referred to as *posterior*. The anteroposterior axis is the main body axis of a typical animal, the others being back-to-front (*dorsal-ventral*) and left-to-right.

ARTHROPOD A *creature* with an external skeleton and jointed legs. The arthropods constitute the largest group of animals, including insects, crustaceans, arachnids, centipedes, and millipedes.

ATHEISM Belief that there is no God, and, usually, no supernatural aspect of things, thus no soul and no afterlife. Sometimes confused with, but quite distinct from, *agnosticism*.

BACTERIUM Simple *creature* consisting of a single *cell* that lacks a *nucleus* and many other internal structures that are found in more complex cells such as those of animals. The word is more often encountered in its plural form, "bacteria." The first cellular life-forms were probably bacteria-like.

BIOSPHERE The "envelope" around Earth in which life-forms are found.

It extends from deep in the soil (or seabed) to a few hundred meters up in the atmosphere.

CAMBRIAN The earliest period of Earth's history (from about 550 to 490 million years ago) to be characterized by abundant and clearly recognizable animal fossils. The contrast between this Cambrian abundance and the earlier paucity of animal fossils has given rise to the idea of a Cambrian explosion of animal life.

CELL The basic building block of all life-forms. Some life-forms, such as bacteria, consist of just a single cell; others, such as animals, consist of many. An adult human consists of about a hundred trillion cells. The number of cell types a life-form is made up of is one measure of its *complexity*.

CHROMOSOME A structure inside the *nucleus* of a *cell* that houses the genes. A typical chromosome looks sausage-shaped when observed with a microscope, and is home to hundreds, or even thousands, of genes.

COMPLEXITY The number of different types of component parts of a life-form. The "parts" can be anything from organs to cells. Evolution has produced a big increase in the average complexity of life-forms in the long term, and *embryogenesis*, or development more generally, produces a big increase in the complexity of the typical animal in the short term.

CREATURE Used in two senses: a broad one, in which it is equivalent to life-form or *organism*; and a narrower one equivalent to animal. Although the word literally means "something that has been created," it is used by both evolutionists and creationists, the former ignoring its literal meaning.

DEVELOPMENT The process of getting from the beginning of a life cycle, usually a fertilized egg, to a reproductively mature adult. In some creatures, such as humans, this occurs directly. In others, such as butterflies, it occurs indirectly, and includes a phase called metamorphosis. In both situations, the earliest stages, occurring within a uterus or an egg case, are referred to as *embryogenesis*.

DEVELOPMENTAL BIAS The tendency of embryos and other developmental stages to be more readily changed in some directions than others by mutations in the genes that govern development. Many consider this phenomenon to be important in determining the direction of evolution, as is Darwinian *natural selection*.

DNA The *molecule* of which the genes are made. Its full name is deoxyribonucleic acid. Its structure, which was discovered in 1953, is that of a double helix in which the two complementary halves of the molecule are wound around each other. DNA is packaged into chromosomes by being supercoiled and coated with proteins.

DORSAL Refers to the back of an animal, or to structures at or near the back. For example, the main nerve cord in a human is dorsal in that it runs down the center of the back (in the spinal column); but in arthropods and mollusks the nerve cord is *ventral*.

EMBRYOGENESIS The making of an embryo. This is the earliest stage of the *development* of an animal, and the most impressive because it involves the production of all sorts of complexities, including *organ* systems, from a starting point of just a single *cell*.

EPIPHENOMENON Something unimportant. A small part of a much larger and more important whole.

EUKARYOTE A type of *cell* that is more complex than a bacterial cell because it possesses numerous internal structures, most importantly the *nucleus*. Also used to describe a *creature* made up of one or more such cells.

EXTANT Still in existence. The opposite of extinct. So humans and horses are extant, while dinosaurs and ammonites are extinct. Can be used to describe either single *species* or groups of related species.

FLAGELLATE Simple *creature* consisting of a single *eukaryote cell* with a hairlike protrusion called a flagellum. The flagellates are thought to be the closest present-day creatures to the ancient unicells that were the ancestors of the animals.

GALAXY A huge aggregation of stars (and their associated planets) separated from other such aggregations by vast areas of empty space. Our own galaxy is called the Milky Way. The nearest other galaxy to us is referred to as "the great galaxy of Andromeda" because it can be seen as a pale blur in the constellation of Andromeda, though the stars that make up that constellation are Milky Way stars.

GENE A functional unit of the *DNA molecule* that makes up a *chromosome*. The typical gene functions by making a *protein* that does some job of work in the *cell* or, more generally, the body. Some genes make proteins that control *development*; others make proteins that have "housekeeping" roles, such as keeping the body supplied with energy.

GENOME The totality of the genetic material within a single *cell* or *organism*. Where an organism is made up of many cells, generally speaking all contain a replicate copy of the same genome, but often with different genes switched on (making protein) and off.

HOMEOSIS A phenomenon in which part of the body develops in a way that is inappropriate for that part but appropriate for another part. Can be caused by a homeotic *mutation* of a *gene* or, in some cases, by exposure to environmental agents that can affect the developmental process. One of the most famous examples is the *development* of legs projecting out of a fly's head, where antennae should be.

IRIDIUM A chemical element, belonging to the platinum group, that is rare on Earth but commoner in some space-derived material, such as meteorites. The presence of unusually high concentrations of iridium in rocks that are about sixty-five million years old has been used to argue that an asteroid impact caused the extinction of the dinosaurs.

JURASSIC A period of geological time, extending from about 200 to 150 million years ago, during which the dinosaurs flourished. Made famous by the film *Jurassic Park*. Followed by the less well known Cretaceous period, at the end of which (65 million years ago) there was a mass extinction.

KINGDOM The largest division of the living world. Originally life was divided just into animal and plant kingdoms. It is now divided into at least five, and often more. The "extra" kingdoms include fungi (previously lumped in with plants) and various kingdoms of unicells, the bacterial kingdom being one of these.

LINEAGE A series of ancestors and their descendants, connected by reproduction through long stretches of evolutionary time. An evolutionary *tree* can be thought of as being made up of many diverging lineages.

MIDDLE WAY A central tenet of Buddhist philosophy, involving a path of moderation and a rejection of extremes. Originally formulated in terms of personal lifestyle, and representing a route between the extremes of opulence and asceticism, it is often now interpreted more widely, as here, where I use it to indicate a route between extreme scientific stances.

MIRACLE Used in two ways. First, to refer to any incredible and wonderful event. Second, to refer specifically to such events when they are thought to be caused by divine intervention in worldly affairs. Here I take the view that the latter, divine sort of miracle does not exist. But there is something miraculous, in the broader sense, in such developmental events as the metamorphosis of a caterpillar into a butterfly.

MOLECULE Just as our bodies are composed of cells, cells are composed of smaller entities called molecules. Unlike cells, these are generally too small to see using an ordinary microscope. They are thus very hard to picture. A useful mental image is to think of a room (the cell) in which someone has exploded several cushions. The room is thus full of feathers (molecules). There is immense size variation among molecules, from the smallest (like water molecules; tiny downy feathers) to *DNA* and proteins (sometimes called macromolecules; peacock tail feathers).

MOLLUSK An animal with a soft body and (often but not always) a shell. Included are snails, slugs, cockles, mussels, octopuses, and squid.

MULTICELLULAR Describes the body of an *organism* that is composed of more than one *cell*. Some very simple multicellular animals are composed of fewer than a hundred cells; more complex ones can be composed of billions or trillions.

MUTATION An accidental change in a *gene*. Mutations can occur in "ordinary" cells of the body (sometimes causing cancer) or in the reproductive cells that will go on to give rise to the next generation. This latter kind of mutation supplies the raw material for evolution. At the molecular level, mutations are alterations in *DNA* sequences.

NATURAL SELECTION The process by which certain variant creatures within a *population* outbreed others and ultimately leave more descendants, thus changing the population's genetic constitution. The driving force, at the population level, behind evolution, as proposed in 1858 by Charles Darwin and Alfred Russel Wallace.

NATURE PHILOSOPHERS A group of early-nineteenth-century European thinkers (mostly German) who adopted a rather idealistic, nonmechanistic view of life. One of the central strands of this view was the *scala naturae*—a vertical axis up and down which creatures could be arranged, with humans at the top. Modern evolutionary theory rejects such a scale and replaces it with a *tree* or bush.

NUCLEUS The body within a *eukaryote cell* within which the genetic material is found in the form of a set of chromosomes, each containing many genes (typically hundreds or thousands).

ORGAN A discrete and sizable unit of structure and function within a body. Examples include the heart, the brain, the eye, and the liver.

ORGANISM An individual *creature* in the broad sense, including those with single-celled bodies and those with *multicellular* ones. Effectively equivalent to "life-form." You and I are both organisms; so are an oak tree and a *bacterium*.

ORGANOGENESIS The production of organs during *development*. Higher animals typically lack any organs when they are early embryos; but later, through multiplication and specialization of cells, they come to possess many organs.

PATTERN FORMATION The process through which organisms become patterned in particular ways during *development*, at levels above that of the individual *cell*. Examples include the formation of muscles of different shapes and the formation of digits of different lengths.

POPULATION All the individuals of a particular *species* living in a particular area. Examples include the human population of Ireland (about five million), the large ground-finch population of one of the Galápagos Islands, and all the individuals of a particular type of *bacterium* living in your gut.

POSTERIOR The tail end of an animal (regardless of whether it actually has a tail); at or near the tail end. Opposite of *anterior*.

PROTEIN A type of *molecule* that is present in, and of much importance to, all cells and organisms. Some proteins make things happen (for example, enzymes); others hold things together (for example, those that combine with fats to make cell membranes). Proteins are made by genes.

RECAPITULATION The phenomenon of the *development* of individual animals supposedly or actually recapitulating stages in the evolution of the animal concerned; an example being the possession of gill clefts by human embryos. There has been much controversy over this idea. It has been both overstated and overly criticized in the past. Basically, recapitulation happens, but in a very imperfect way, and not always—that is, it does not have the status of a law.

SCALA NATURAE Translates as "natural scale." An old concept—the arrangement of all creatures on a single scale of advancedness—pursued by the nineteenth-century *nature philosophers*. It is now recognized that the diversity of life-forms is far too great for a single such scale to be meaningful.

SEGMENT One of a series of structural units repeated, with more or less variation, along the anteroposterior body axis of many animals, including vertebrates, arthropods, and some kinds of worms (for example, earthworms and ragworms).

SPECIES A group of creatures that can interbreed with each other but that are reproductively isolated from others. Animal species are often quite clear-cut; examples include our own, human, species; also such familiar ones as the gray squirrel and the starling. Species are generally harder to delineate in the other kingdoms.

THEISM Belief in the existence of God (or indeed of gods). From the Greek *theos*, meaning "god." Opposite of *atheism*. The world's main religions are different forms of theism. Christianity, Judaism, and Islam are all monotheistic religions—that is, they are based on a belief in the existence of only one God. (Of course, all this begs the question of who or what God is.)

TISSUE A sheet or block of cells, generally of the same broad type. So, for example, we speak of skin, muscle, and nervous tissue. An *organ*, in contrast, is usually made up of several different tissues interspersed with each other.

TRANSCRIPTION (FACTOR) When a *gene* is switched on, and hence making its product, it is said to be being transcribed. Actually, gene function is (at least) a two-stage process: to make a *protein*, a gene undergoes transcription to make an intermediate product from which a protein is ultimately made by the second stage of the process (called translation). Although many proteins remain outside the *nucleus* and do all sorts of jobs in the *cell* "at large," some enter the nucleus and switch genes on or off. These special proteins are called transcription factors.

TREE (EVOLUTIONARY) A picture of the treelike radiation of different ancestor/descendant lineages as they go their separate ways from the starting point of a common ancestor. When there is particularly rapid radiation of many lineages, a tree is said to be bushlike.

UNICELL A *creature* whose body consists of just a single *cell*. Examples include bacteria, amoebae, and flagellates.

VENTRAL Refers to the front of an animal, in the sense that it is opposite to the back. However, the actual orientation of the ventral surface depends on posture. So in humans, ventral really is frontal, whereas in a more typical four-legged mammal such as a dog, the ventral surface is the underside. Opposite of *dorsal*.

VERTEBRATE Any animal with a backbone composed of vertebrae. The groups of vertebrate animals are mammals, birds, reptiles, amphibians, and fish. Together they include some fifty thousand *species*.

WORLDVIEW An overall philosophical view of the nature of things. Not to be interpreted as specifically a view of Earth. Rather, it refers to someone's view of "life, the universe and everything," to borrow again from Douglas Adams. A worldview is likely to include, as one of its main components, *theism* or *atheism* or *agnosticism*.

FURTHER READING

I give here ten suggestions for further reading accessible to a general audience. In all cases, check for more recent/cheaper editions!

First, the ultimate accessible classic:

Darwin, C. 1982. *On the Origin of Species by Means of Natural Selection; or, The Preservation of Favoured Races in the Struggle for Life*. Penguin, London. (Original edition published by John Murray, London, 1859.)

Second, a highly readable account of the history of life, that is, what has happened on our planet, in evolutionary terms, over the last four billion years:

Fortey, R. 1997. *Life: An Unauthorised Biography*. HarperCollins, London.

Next, a story about how that other creative biological process, development, works, with special reference to the role of genes:

Coen, E. 1999. *The Art of Genes: How Organisms Make Themselves*. Oxford University Press, Oxford and New York.

Joining evolution and development together:

Carroll, S. 2005. *Endless Forms Most Beautiful: The New Science of Evo Devo*. W. W. Norton, New York.

Specifically on the evolution of complexity, that is, my main theme here, but with a somewhat different approach:

Maynard Smith, J., and E. Szathmáry. 1999. *The Origins of Life: From the Birth of Life to the Origin of Language*. Oxford University Press, Oxford and New York.

On the "science of complexity," an often-hard-to-penetrate field, but here made more accessible than usual:

Kauffman, S. 1995. *At Home in the Universe: The Search for Laws of Self-Organization and Complexity*. Viking Penguin, New York and London.

And finally, on the religious implications of evolutionary theory, here are three different perspectives, respectively theistic, atheistic, and agnostic, the last of these being historical, and relating to the original agnostic, Thomas Henry Huxley:

Conway Morris, S. 2003. *Life's Solution: Inevitable Humans in a Lonely Universe.* Cambridge University Press, Cambridge and New York.

Dawkins, R. 1986. *The Blind Watchmaker.* Longman, London.

Desmond, A. 1994 and 1997. *Huxley: The Devil's Disciple* and *Huxley: Evolution's High Priest.* A biography of Thomas Henry Huxley in two volumes. Michael Joseph, London.

ACKNOWLEDGMENTS

My main aim here has been to write as clear, accessible, and brief a book as possible while doing the subject matter—life—justice, and without compromising scientific rigor.

Many people have helped me in my attempt to achieve this aim, by reading the entire manuscript, or at least a substantial chunk thereof. These folk range from academic and publishing professionals to friends and family. It is now my pleasure to thank all of them for their help while acknowledging, in the time-honored way, that I claim complete responsibility for any failings of clarity, accessibility, brevity, or rigor that remain.

So a very big thank-you to the following: Ward Cooper, Ed Knappman, Alec Panchen, Mary Scott, and Joe Wisnovsky; plus a plethora of Arthurs, ranging from students to fifty-somethings: Stephen, Michael, Claire, Helen, and Chris.

INDEX

Acrania, 188
African Genesis (Ardrey), 195–96
agnosticism, 231–32
albinism, 127, 128, 131
alphabet, genetic, 152, 155
amino acids, 51–53
amoebae, 162
amphibians, 189, 212; *see also* frogs
Andromeda, great galaxy of, 221
anemia, sickle-cell, 51–53
angiosperms, 98
angularity, 30–31
antennapedia, 85, 154
Anthropogenie (Haeckel), 107
ants, social castes of, 78
applied science, 20–21
archaea, 162
Ardrey, Robert, 195–96
Aristotle, 68
Armstrong, Neil, 121
arthropods: feeding behavior of, 143–44; Hox genes of, 159; segmental structure of, 139–40, 148, 153–54,
Ascent of Man, The (Bronowski), 8
asexual reproduction, 71, 133
Asimov, Isaac, 218
asteroid impact theory of mass extinction, 209–11
atheism, 230–32
atomic bomb, 21
Australia, 98
average complexity, 98, 121–22, 179–80, 182
bacteria, 4–5, 97, 119–23, 132, 158, 162; cells of, 57, 91, 92, 119–20; complexity of humans versus, 11; filamentous designs of, 8; reproduction of, 37
Baer, Karl von, 101, 104–106, 108–11
balancing selection, 52–53
barnacles, 33
basalt, 210–11
Bateson, William, 153–54
Beagle (ship), 47, 108
bees, social castes of, 78
behavioral complexity, 10, 172, 193–203, 213, 215
Biased Embryos and Evolution (Arthur), 131
bicoid protein, 84
big bang, 6, 227, 228
bilaterality, 30, 159, 173, 179–82; cephalization and, 189; departures from, 180–81
bilateral symmetry, 30
biogenetic law, 105, 107
biospheres, 12, 215, 222; evolutionary events on scale of, *see* mass extinctions
bipedalism, 196
birds, 239, 189, 212–14; beaks of, 45–47, 119; brain size in, 213; embryos of, 110; lineage of, 94; size variation in, 181; wings of, 142
bithorax, 86
Blind Watchmaker, The (Dawkins), 50, 138, 230
blood cells, 59, 82–83; *see also* hemoglobin

body forms: complexity of, 172, 174, 179; of jellyfish, 169–71, 173; of multicellular creatures, 164–65; of sponges, 167, 173; *see also* symmetry
Bonner, John Tyler, 41
boxes, in molecular genetics, 152, 153; *see also* homeobox
brain, 172; cells of, 60; development of, 190–93, 202–203; evolution of, 186–90, 194–99; human, capacities of, 194, 199–203; octopus, 141, 147–49, 176, 192, 201, 202; size and complexity of, 182, 193, 194, 197–98, 213, 215
British Columbia, 96
Bronowski, Jacob, 8
Brown, Christy, 177
Buddhism, 6, 15, 22
Burgess Shale, 96, 97
butterflies, 11, 110–11; complexity of, 160, 171; metamorphosis of, 76–77, 111

Cambrian explosion, 173, 195, 218
Cambrian period, 95–99, 208
Canada, 96
Canidae, 45
carbon-based life forms, 119, 220–21
case studies, 44–54, 118
cells, 55–66, 70, 79–80, 161, 221; blood, 59 (*see also* hemoglobin); complexity and types of, 11; death of, 77; in developmental processes, 71–77; differentiation of, *see* differentiation, cellular; division of, 56, 60–61, 64, 71, 73, 74, 76, 157; eukaryotic, 92–93, 97, 161; genetic material of, *see* genes; of jellyfish, 168; movement of, 62–65; nerve, 148–49, 168, 175–76; origin of, 38; prokaryotic, 97; properties of, 59–60; proto-, 91; reproduction of, 35–39; simple, 92, 93, 119; of sponges, 167; sticky, *see* "Velcro," intercellular; structures of, 57; types of, in stem lineages, 49, 132–33
cell theory, 57–58, 67
Cenozoic era, 208, 210
centipedes, 131, 144–46, 148, 184, 201; complexity of, 160
cephalization, 185–93; development of, 190–93; evolution of, 185–90, 192–93
cerebrum, 199, 213
chicks, descriptive studies of development of, 68
Child, C. M., 84
chimpanzees, 195, 198; behavioral capability of, 199–200, 213; consciousness of, 201
China, 96; ancient, 7
choanocytes, 167
Christianity, 226, 228–30
chromosomes, 87
ciliates, 162
circular reasoning, 35
cleavage, 71–72, 74, 75
Cnidaria, 171
convergent evolution, 187, 214
co-option, 178–79, 190, 191
coal-age forests, 145–46
collar cells, 167
communication: intercellular, 63–64; of science, 114–18
comparative developmental biology, 68, 84
complexity, 5, 6, 8, 10–13, 15, 25, 172–

82, 227–28; ascents in, 40–45, 55; of brain, 182, 193, 194, 197–203; case studies and, 44–54; compromise point of diversity and, 19; definition of, 11; developmental increases in, 79; through diversification of replicated parts, 136, 149; effect of mass extinctions on, 205, 207, 211–15; embryonic trajectory of, 108, 112; genetic, 150–60; intelligent-design claims about, 229; of life cycles, 171; measures of, 12; natural selection and, 118–25; punctuated equilibrium and, 51; scientific interest in, 26; steps from simplicity to, 161–71; over time, 12–14, 92–100; types of, 10–11
compromise, 15–17; between lawn and ladder views of life, 17–19; Middle Way as argument for, 15
computer viruses, 34
concentration gradients, 84
consciousness, 201–203, 228
Copernicus, Nicolaus, 7
corals, 171
cranium, 188
creationism, 6, 7, 116, 192, 226–30
Crick, Francis, 13, 40, 51, 67, 151
crocodiles, 212; lineage of, 94
crosses, 150–51
Crusades, 228, 232
crustaceans, 185; fossil, 96
cytoplasm, 57, 63–64, 120; egg cell, 83

Darwin, Charles, ix, 7, 8, 24–25, 28, 42, 102, 105, 117–18, 179; on embryology, 108, 109, 112; finches studied by, 47–48; Huxley and, 116, 231; legacy of, 125; natural selection theory of, *see* natural selection; Wallace and, 13, 114–17
Dawkins, Richard, 50, 138, 192, 230
death, cellular, 77
de Bono, Edward, 201
deoxyribonucleic acid, *see* DNA
Descent of Man, The (Darwin), 8
descriptive developmental biology, 68, 84
deuterostomes ("mouth-seconds"), 73, 94, 96
development, 56, 67–78, 100; cellular processes in, 62; general theory of, 68–69, 81, 84; genetic aspects of, 65, 69–70, 79–88; of human brain, 202–203; increases in complexity in, 99; nature philosophers' view of, 101–104; processes of, 71–77; reprogramming of, 128–33, 146, 179, 190; von Baerian, 104–105; *see also* embryogenesis
developmental bias, 131, 134–35
differentiation, cellular, 56, 61–62, 73–75; genes and, 82–84, 86
dinosaurs, 216, 218; extinction of, 25, 195, 206, 208, 209, 212–14; feathered, 94
divergence, von Baerian, 104–105, 107, 109–11; *see also* duplication and divergence
diversity, 5, 6, 15, 40–41, 227–28; compromise point of complexity and, 19; in case studies, 45–49, 53
division, cellular, 56, 60–62, 64, 70–71, 74, 76, 157
DNA, 40, 51, 61, 87, 133–34, 151–57; divergence of sequences of, 93–94, 96

dogs, 45; consciousness of, 201
Dolly (cloned sheep), 83
duplication and divergence, 157, 165, 179; genetic, 150–60, 164, 179, 190; structural, 141–49, 158

earthworms, 131, 174; regeneration in, 183
echinoderms, 181
ectoderm, 169
Ediacaran biota, 98–99
egg, 59, 70, 151; cytoplasm of, 83; fertilized, 71, 99–100, 109, 129, 130, 190; proteins of, 84
Einstein, Albert, 21, 221
Eldredge, Niles, 50
electron microscopes, 56
embryogenesis, 67, 69, 70, 88, 94, 101, 108, 139–40, 147, 151; cellular processes in, 61, 62, 65, 77; of cephalization, 188, 190–93; cleavage in, 71–72, 74, 75; consciousness and, 203; gastrulation in, 72–75; mutation in, 129–34, organogenesis in, 74–76; proteins in, 84; transcription factors in, 155
embryology, 169; comparative, 102, 104–12; complexity and, 112–13; experimental, 68; nature philosophers' view of, 101–105
endoderm, 169
environmental variation, 127–28, 130, 140, 206
enzymes, 61, 119, 155
epiphenomenon, 43; humans as, 4, 5, 18, 41, 70
ethos, scientific, 21
eukaryotic cells, 92, 97, 121–23, 161, 215
Evolution and Development (journal), 70
Evolution of Man, The (Haeckel), 107
experimental developmental biology, 68, 84
experiments, ideal, 116–17, 221
explanation, generality of, 21–23, 25
extinction, 23–25, 124, 136, 138; background rate of, 206; of dinosaurs, 25, 195, 206, 208, 209, 212–14; fossil evidence of, 96; habitat destruction as cause of, 204; inevitability of, 32; of protohuman side branches, 196; *see also* mass extinction
extraterrestrial life, 216–25
eyes, 187; color of, 129, 130, 132

Fabricius, Hieronymus, 68
fascism, intellectual antecedents of, 113
filamentous designs, 8
finches, 47–48, 118
fish, 142, 168, 189; embryos of, 101, 110; eyeless, 139; Hox genes of, 159; jawless, 198, 189
fitness, 118, 125, 127, 143, 207; variation and, 128, 134; *see also* natural selection
flagellates, 162, 167
flatworms, 159, 174–76, 185; regeneration in, 183
flowering plants, 98, 145–46
forward movement, 175–76
fossils, 91, 145, 206, 208; and age of lineage divergences, 93–94; Cambrian, 95–96, 173; case studies based on, 45, 48–51; Ediacaran, 98–99; protohuman, 195, 197–98
fraud, scientific, 111–12, 226

French Revolution, 183
frogs, 171; developmental research on, 69–70, 83, 87, 110; overproduction of, 137
fruit flies, 83–87, 221; isogenic lines of, 127; mutations in, 85–86, 138–39
fundamentalism, religious, 228–30
fungi, multicellular, 98

Galápagos Islands, 47–48, 108
galaxies, 220–21
Galileo Galilei, 27, 230
ganglia, 148, 176
gastrulation, 72–74
generality, 21–23, 25; of relationship, 34–35
genes, 52–53, 57, 65, 92, 102, 119, 150–60, 206; altered, 125; co-option of, 178–79; development and, 67–69, 79–88, 151; duplication of, 60–61; homeobox, 68, 153–59; Hox, 159–60; molecular studies of, 54; mutations in, *see* mutations; switching on and off of, 61–62, 64, 74, 80–83, 129, 130, 139, 156, 166, 191; variation in, *see* variation, genetic
genomes, 79, 83; complexity of, 158–59; of organelles, 92
geology, 95
German nature philosophers, *see* nature philosophy
gills, 141–42; embryonic, 73, 101, 103; rudimentary, 177–78
global warming, 203, 209
globins, 158
Gould, Stephen Jay, 17, 40, 50, 70, 112, 113, 131, 217
gradualists, 192
greenflies, 78
growth, 76
Gurdon, John, 83, 87
gut, development of, 72–73

habitat destruction, 203–204, 209
Haeckel, Ernst, 42–43, 101, 105–109, 111
Harvard University, 50
heads, 182, 183–85; origin of, 25, 175–76, 179; *see also* cephalization
heliocentric solar system, 7, 16
hemoglobin, 51–53, 82–83, 155, 156, 158, 205
Hendrix, Jimi, 119
Hennig, Willi, 13
heterozygotes, 52–53
homeobox, 68, 153–59
homeosis, 154
homeotic mutations, 86
Homo sapiens, 90, 199
homology, 187
homozygotes, 53
Hooke, Robert, 56–57
Hooper, Judith, 46
horses, size variation in, 48–50, 181
Hox genes, 159–60
Human Genome Project, 61, 79
humanoids, 195–97, 203
hummingbirds, 181–82
Huxley, Thomas Henry, 116, 231–32
hydras, 171
hydrochloric acid, 59

ideal experiments, 116–17, 221
imperialism, genetic, 65–66, 80

Indian subcontinent, 210, 211
industrial melanism, 45–47, 53, 120, 128, 205–207
inert substances, 36, 38
inheritance, Mendelian laws of, 67, 111, 150–51
insects, 217; embryos of, 110; heads of, 185; *see also specific insects*
instinctual behavior, learning versus, 10
intelligence, 172
intelligent design (ID), 6–7, 229
interest, scientific, 19–22; of origins, 24; of reproduction, 35
intermediate structures, usefulness of, 142
intuition, scientific, 25–26
Iraq, 172
iridium, 209–11
isogenic lines, 127

jellyfish, 161, 166, 168–71, 173–76, 181, 185, 201, 212; box, 31
Jensen, Zacharias, 56
Jesus, 228, 229
Judaism, 228, 230
Jurassic period, 214

Kennedy, John F., 106
keratin, 155
Kubrick, Stanley, 196

ladder view of life, 5–6, 8, 10–11, 26, 41; compromise between lawn view and, 17–19, 22; emphasis on differences in, 10; ideological taint of, 41–42; as mechanistically explicable, 42–43; scientific interest of, 25
lampreys, 188
lancelets, 188–89, 192
language, capacity for, 194, 200
larvae, 11, 110; of sponge, 167
lawn view of life, 5–6, 10–11, 41, 43; compromise between ladder view and, 17–19, 22; scientific interest of, 25
learning, capacity for, 10, 199–200, 213
Lewis, Ed, 68–69, 87, 154
Li, Jet, 226
life, origin of, 24
life cycle, complexity of, 10–11
"life essentials," 32
limpets, 33
lineages, 8, 13, 91, 215; age of divergences in, 93–94; changes in complexity in, 97, 122; comparative embryology and, 108–109; extinct, 138; humanoid, 90, 195–97; occurrence of change in, 24; of parasites, 139; particular, studies of, 44; punctuated equilibrium in, 50; of red blood cells, 82; segment increases and decreases in, 140; splits in, 13–14, 24, 124, 187–88, 195; types of cells in, 48–49, 132–33
Linnaeus, Carolus, 115
Linnean Society, 114, 115
lizards, 212, 213

MacArthur, Robert, 23
macroevolution, 49, 51, 54, 206
malaria, resistance to, 52–53
mammals, 124, 176–77, 189, 212–14, 216; bilateral symmetry of, 180;

brain size in, 213; cardiovascular system of, 178; diversification of, 195; embryos of, 110; size variations in, 181; *see also specific mammals*
mass extinction, 123, 204–17; causes of, 209–11; effect on complexity of, 211–15; habitat destruction and, 204; percentage of species killed off in, 208–209; survival of the luckiest in, 207
Materials for the Study of Variation (Bateson), 153
Matrix, The (movie), 109
maximum complexity, 98, 121, 123, 158
Maynard Smith, John, 32, 41, 229
Mechanism of Mind, The (de Bono), 201
Meckel, Johann, 101, 102, 109, 113
medusa body form, 171
megaevolution, 49–51, 54
melanism, industrial, 45–47, 53, 120, 128, 205–207
membrane, cellular, 57, 62, 92, 161, 163
Mendel, Gregor, 67, 111, 150–51, 153
mesoderm, 169
Mesopotamia, 172
Mesozoic era, 208, 210, 211, 214
metamorphosis, 76–77, 88, 111, 137
microbes, 161; *see also* bacteria
microevolution, 47, 49, 53, 54, 206
microscopes, 56–57
Middle Way, 6, 12, 15, 19, 43–44; biased, 22, 39, 43–44
Milky Way galaxy, 221
miracles, 228
mobility, 33–35
modular design, 140, 146, 148
molecules, 119; of cells, 37, 56, 63–64; hemoglobin, 51–52, 64; large, aggregates of, 38, 91, 93, 123, 161, 227; protein, *see* proteins; sticky, 163–64; studies of, 53–54
mollusks, 12, 45, 217; brains of, 148–49, 176; departures from bilaterality in, 180; fossil, 50–51, 96; heads of, 185; shells of, 35–36, 38
moths, industrial melanism in, 45–47, 53, 120, 128, 205–207
movement: cellular, 56, 62–65, 73, 74, 76; forward, body form and, 175–76
multicellular creatures, 98, 179, 218, 221; body forms of, 164–65; diversification of, 96; life cycles of, 166; origin of, 24, 92, 93, 96, 97, 161–64
multiverse, 226
Muslims, 228–30
mutations, 78, 125, 137–38, 143, 179, 205, 215; duplication and, 138–39, 140, 142, 164; homeotic, 154, 155; in pattern formation, 85–86; variation produced by, 128–34

natural selection, 8, 78, 88, 105, 114, 137–38, 179, 205, 222, 231; balancing of mortalities and, 52–53; body form and, 175; brain size and, 202; cephalization and, 190–92; comparative embryology and, 108–109; complexity and, 118–25; directional, 53, 131; life cycles and, 91; in moths, 46; micro-level, 38–39; mutation and, 78; overemphasis on, 125; punctuated equilibrium and,

natural selection (*continued*)
50; size and, 177; theory of, 67; variation and, 126–35
Nature (journal), 51
nature philosophy, 5, 18, 19, 42, 213; comparative developmental theory in, 68; embryology in, 101–13
Neanderthals, 196
nerve networks, 168–70, 175–76, 181
Newton, Isaac, 221
notochord, 188
nucleotides, 61
nucleus, 57, 92

Occam's razor, 17, 220
octopus, brain of, 141, 147–49, 176, 192, 201, 202
Of Moths and Men (Hooper), 46
Ohno, Susumu, 135
Oken, Lorenz, 113
One, The (movie), 226
On the Origin of Species (Darwin), 7, 24, 28, 102, 108, 112, 116–17, 179
organelles, 57, 92
organisms, 150, 151; development of, *see* development; life cycles of, 11; reproduction of, 37; static, 33
organogenesis, 74–76
organs, 11, 55–56; asymmetrical, 180; complexity of, 11, 177–79; differentiation of, 18
origins, 25; interest in, 24; of cells, 38

pair-bondings, 126
paleontology, 48, 50, 95, 206, 207; *see also* fossils
Paleozoic era, 208, 211, 214
Paley, William, 229
parasites, 10–11, 59, 139, 213
parsimony, principle of, 17
pattern, process versus, 22
pattern formation, 75–76, 80, 87; mutation and, 85–86
periods, geological, 95
philosophical baggage, 9–10, 14, 19, 22; conflation of time and complexity in, 13; genocentric, 65–66, 81, 130; politically based, 42–43; recapitulationist, 113; scientific intuition versus, 25–26
photosynthesis, 92, 146
plants, 33, 98, 182; Ediacaran, 99; flowering, 145–46; intraspecies differences in, 78; modular increases and decreases in, 140; photosynthesizing, 92, 146
plasma, 59
polyp body form, 171
population, 12
Porifera, 170–71
Portuguese man-of-war, 168, 176
positional information, 84
primates, 195
primordial soup, 38, 91, 227
process, pattern versus, 22
prokaryotic cells, 97
proteins, 61, 65, 151, 152; blood, case study of, 45, 51–53; boxes and, 153, 155; of cell membranes, 163; developmental role of, 80–84, 139; gene-switcher, 155–57; in intercellular communication, 63–65
proto-cells, 38, 123
protohumans, 90
protostomes ("mouth-firsts"), 73, 96
protozoans, 162, 167

punctuated equilibrium, 50–51
pure science, 21; generality in, 23

racism, 43
radial cleavage, 71
radial symmetry, 169
radiation, post-extinction, 212
range of complexity, 98
reasoning, circular, 35
recapitulation, 105, 107–109, 113, 190
rectilinearity, 31
redshifts, 227
redundancy, 140, 142, 164
regeneration, 88, 183
regularity, 30
relatedness, patterns of, 99
relativity theory, 58
religion, relation of science and, 225–32
replication, 139, 221; diversification and, 142–49, 158–60, 179
reproduction, 32–35, 161, 165; asexual, 71, 133; of bacteria, 4; of cells, 35–38; non-variant, 126; of sponges, 167
reptiles, 212; brain size in, 213; *see also* dinosaurs
Richardson, Michael, 111
Robert, Jason Scott, 17
roundworms, 131, 159
Roux, Wilhelm, 68
Royal Society, 115
Rushdie, Salman, 230

Saint-Hilaire, Étienne Geoffroy, 189
saltations, 51, 143
sand, nature of, 35–37
Sander, Klaus, 110
scala naturae (natural scale), 6, 41–42, 102
scale, 30; angularity and, 31; of reproduction, 37
Schleiden, Matthias, 57–58
Schwann, Theodor, 57–58
Scott, Andrew, 123
sea anemones, 171, 175
sea cucumbers, 181
sea stars, *see* starfish
sea urchins, 110–11, 181
segmentation, 139–40, 148, 153–55, 174
segment identity, 84–85, 87
selector genes, 86, 87
Selfish Gene, The (Dawkins), 192
sequence similarity, 52
sexes, different developmental outcomes in, 78
sexual maturation, 74
shapes, 30; of cells, 59; *see also* body forms
Siberia, 211
sickle-cell anemia, 51–53
simplicity, scientific search for, 16–17
Simpson, G. G., 44, 48, 49, 51
simultaneity, 223
size: brain, 182, 193, 194, 197–98, 213, 215; increase in, 176–77, 179–81; variation in, 181–82
slime molds, 58
sophistication, *see* complexity
solar systems, 218–21, 227; as heliocentric, 7, 16
Soviet Union, 232
speciation, 23–24
species: of bacteria, 4; belief in immutability of, 24; differences

species (*continued*)
within, 78; evolution within versus between, 49; extinction of, *see* extinction; facts of development in, 69; of finches, 47–48; of moths, 45; number of, 23; overproduction of, 137; reproduction of, 32, 34; size variation in, 181
spiral cleavage, 71
sponges, 159, 166–71, 173, 201, 212
Stalinism, 232
starfish, 33, 181, 186
stem lineage, 48–49
structural complexity, 10
symbiosis, cellular, 92, 93
symmetry: bilateral, *see* bilaterality; radial, 169, 173–75, 179, 181, 186
syncytium, 58–59
synthesis, 44
Szathmáry, Eörs, 41

tapeworms, 10–11, 139
TATA box, 153
tautologies, 35
taxonomic scope, 49
temperature, extremes of, 219
tempo of evolution, 51
testable theories, 17
theism, 231; *see also* religion
time: of cellular life span, 59–60; complexity and, 12–14; differentiation over, 18; of evolutionary change, 45–46, 49; geological, 208; life through, 89–100, 165–66; linearity of, 42
tissues, 55–56, 74
transcription factor, 155–57
trees, evolutionary, 13, 14, 22, 186, 202; directionality of, 8–9
trilobites, 209
truth, economy in search for, 17
Turkana, Lake, 50
Turner, Brian, 80
turtles, 143
2001: A Space Odyssey (movie), 196

unicells, 11–12, 88, 93, 161–62, 179, 218–19, 227; ancestral, 38; diversity of, 12; fertilized eggs as, 100; fossil evidence of, 94; origin of multicellular creatures from, 24; reproduction of, 37
universe, 227–29; age of, 90; birth of, 6, 227; expanding, 227; multiple, 226; structure of, 220–21
Ur, ancient city of, 172
urbilaterians, 173–77, 185, 186
Ussher, James, 231

variable gene activity theory of cell differentiation, 62
variation, 126–35, 140; cephalization and, 190; discrete, 153–54; in size, 181–82
"Velcro," intercellular, 56, 62–64, 163–64, 215; in embryogenesis, 73
vertebrates, 12, 69–70, 148, 170, 195, 217; cephalization of, 185–90; comparative embryology of, 110, 111; complexity of, 98; fossil, 96; front end of, 141–43; Hox genes of, 159; mass extinctions of, 212; organogenesis in, 74; origin of, 24; segment increases and decreases in, 140

viruses, 58, 123; computer, 34
vitalism, 103–104
volcanic-activity theory of mass extinction, 209–11

Waddington, C. H., 53
Wales, 95
Wallace, Alfred Russel, ix, 13, 114–17
warm-bloodedness, 213–14
water bears, 219
Watson, James, 13, 40, 51, 67, 151
Whitehead, Alfred North, 16
Wilberforce, Samuel, 116
William of Occam, 17
Williamson, Peter, 50
Wolpert, Lewis, 70, 84
worldviews: convergence-followed-by-divergence pattern and, 110, 111; eighteenth-century, 7; evolutionary, 6–7; relationship between science and religion in, 225–32; threats to, 26
worms, 10, 131, 159, 174, 183, 238; fossil, 96; parasitic, 10–11, 59, 139

yeasts, 162